CLASSIC GUNS OF THE WORLD SERIES

THE COLT M1911

.45 Automatic Pistol

OTHER BOOKS IN THE SERIES

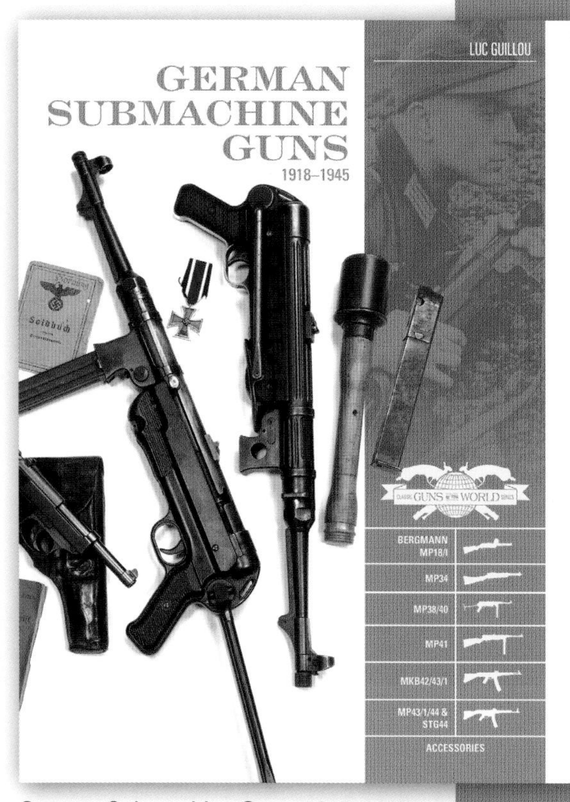

German Submachine Guns, 1918–1945
978-0-7643-5486-1

American Submachine Guns, 1919–1950
978-0-7643-5484-7

CLASSIC GUNS OF THE WORLD SERIES

The Sten
978-0-7643-5485-4

The Luger P.08 Vol. 1
978-0-7643-5657-5

JEAN HUON
THE COLT M1911
.45 Automatic Pistol
CLASSIC GUNS OF THE WORLD SERIES
M1911
M1911A1
MARKINGS
VARIANTS
AMMUNITION
ACCESSORIES
MILITARY
4880 Lower Valley Road
Atglen, PA 19310

Originally published as *Le Colt .45: Le pistolet de J. M. Browning* by Regi'Arm, Paris © 1999 Regi'Arm
Translated from the French by Julia and Frédéric Finel

Library of Congress Control Number: 2019935941

Cover design by Justin Watkinson
Type set in Helvetica Neue LT Pro/Times New Roman

ISBN: 978-0-7643-5825-8
Printed in China
5 4 3 2

Published by Schiffer Publishing, Ltd.
4880 Lower Valley Road
Atglen, PA 19310
Phone: (610) 593-1777; Fax: (610) 593-2002
E-mail: Info@schifferbooks.com
Web: www.schifferbooks.com

CONTENTS

INTRODUCTION

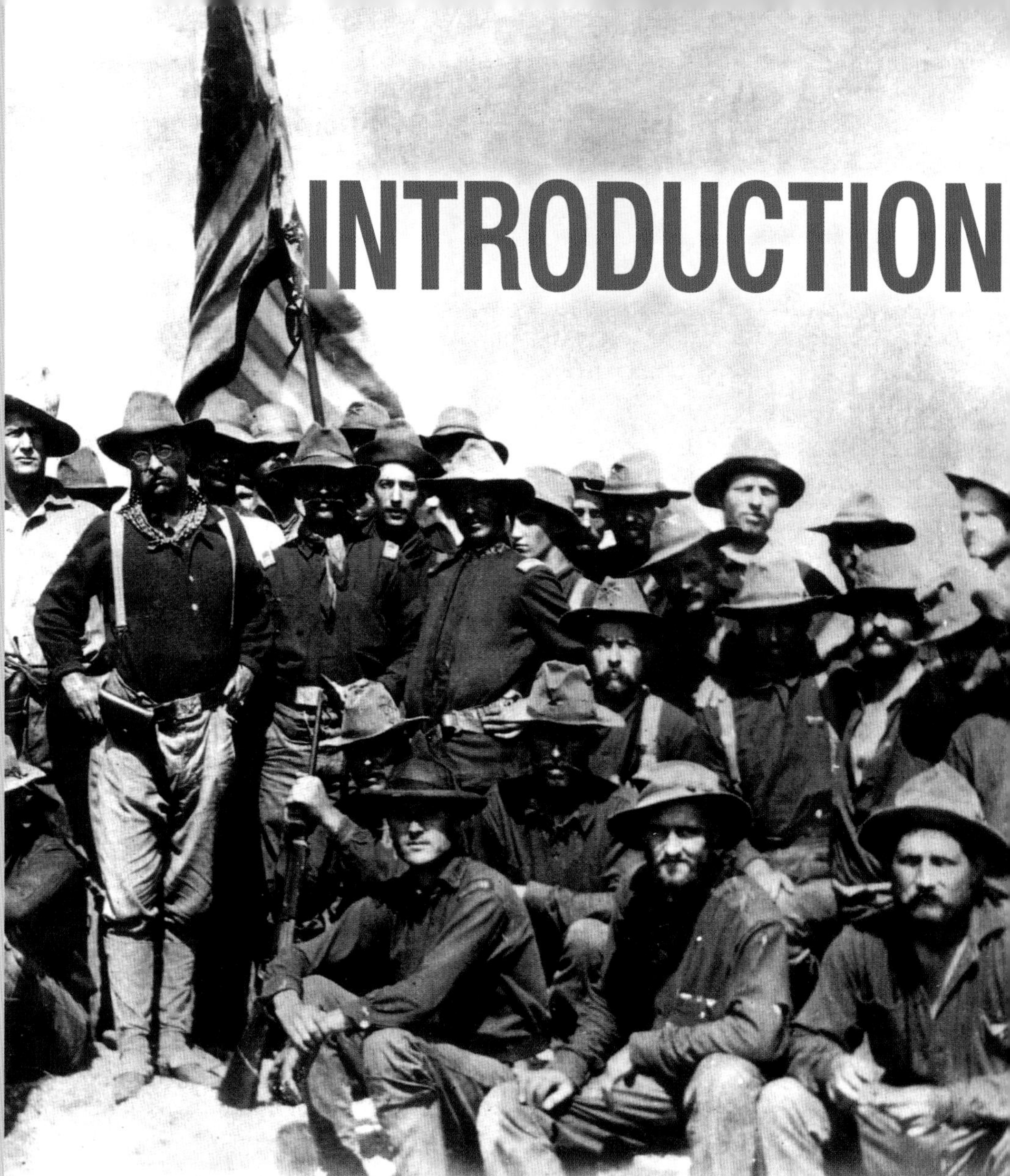

Lt. Col. Theodore Roosevelt, future president of the United States, and his men of the first cavalry of volunteers in 1898 during the Spanish-American war. *HGPC*

Soldier of the US Cavalry during the 1880s, holding his regulation Colt single-action revolver

More than in any other army, the handgun held a dominant position in the arsenal of the American soldier.

From 1847 onward, under the drive of Capt. Walker, cavalry units were equipped with revolvers. It was still a percussion-type revolver, both heavy and bulky, but its .44 (11 mm) caliber ensured its great power.

The official doctrine of the use of the cavalry in the American army was entirely dissimilar to what was the case in Europe at the same time. In Europe the light cavalry was used for reconnaissance missions and surprise attacks, whereas the heavy cavalry intervened to dismantle the enemy infantry square.

In the United States, only one type of cavalry existed and was often associated with infantry units and light artillery batteries. In the nineteenth century, the enemy was the "Red Indian," and the cavalry that carried out missions to protect the territories of the West derived its effectiveness from its mobility and firepower. The Civil War bore little influence on this tactic, but it boosted the development of new technologies and industry. From then on, weapons firing metal cartridges were used, and in 1873 the Colt .45 six-shot revolver, fitted with a single-action trigger, became part of infantry, cavalry, and artillery equipment.

In 1889, the US Navy, followed by the US Army in 1892, adopted a Colt revolver with a simultaneous extraction of all shells, firing a .38-caliber cartridge. The Philippines campaign was to reveal the lack of power of the ammunition, and in 1902 the army returned to the .45-caliber revolver.

It was around this time that the military became interested in the development of semiautomatic pistols. The choice of the new weapon was not going to be easy, particularly since it involved satisfying the wishes of the infantry with those of the cavalry.

CHAPTER 1

JOHN MOSES BROWNING

The Browning brothers' gunsmiths store in Ogden. *Fabrique Nationale*

John Browning, toward the end of his life, with the famous shotgun that bears his name. *FN*

John Moses Browning was born on January 23, 1855, in Ogden, Utah, a small town situated at the foot of the Rocky Mountains by Salt Lake. His father, Jonathan, had settled in the town a few years earlier after a difficult time and converting to the Mormon religion. Jonathan Browning, a qualified gunsmith, set up a small workshop in the region, which was not only still hostile at that time but also devoid of any industrial infrastructure. He worked as a gunsmith, wheelwright, and general mechanic. This meant he could make flintlock and then percussion-type rifles and, most notably, a shotgun fitted with a "harmonica" magazine. Very soon his second son, John, showed he possessed great manual dexterity and phenomenal ingenuity; he was also a good horse rider, a formidable hunter, and a fine shot. With his brothers, Matt and Ed, he was not short of creative ideas.

In 1870, at a time when Salt Lake City and Ogden were linked by railroad, their father handed over the business to them, and in 1878 the Browning brothers created the firm that still bears their name today. Their main activity was repairing weapons, but they also created and made a new single-shot carbine. The company grew and prospered, and in 1880 they built a workshop equipped with a 5 hp steam engine. However, production was not able to follow the ideas conceived by the fertile mind of John Moses. Fortunately, some large companies became interested in his creations, first of which was Winchester, with its standard carbine and several lever-action carbines.

In 1890, he became interested in automatic weapons and made prototypes of rifles and machine guns. But although Winchester bought the majority of his inventions (he made up to three per month), it did not produce them all, which led the inventor to propose them elsewhere. In 1895, Colt produced the first Browning-system machine gun, adopted by the American armed forces.

TOP: one of the first 1900 models. **MIDDLE:** the serial number 1 on the 1900 model. **BOTTOM:** the luxury 1903 Hammerless .32-caliber model.

A 1909 Colt advertisement showing the range of products

After rifles and machine guns, he became interested in pistols. His first prototypes used very diverse mechanisms, whose applications interested not only Colt but also the "Fabrique Nationale" in Belgium, with whom Colt collaborated on a Long Rifle .22 carbine and a hunting rifle, both automatic.

But in reality, it was with Colt that its combat pistol was developed and taken up by the US Army in 1911.

Final checking in the Colt workshop around 1926

During the First World War, Colt contributed to the war effort by developing standard machine guns, light machine guns, a heavy machine gun, and a 37 mm automatic barrel. When peace returned, the company came back to Belgium, where Browning again developed a hunting rifle along with a prototype of what was to become the G.P. 35 pistol.

John Browning, the creator of several hundred patents and the greatest weapons inventor of our time, died in Herstal on November 26, 1926.

CHAPTER 2

HISTORY

Description of the Browning short-recoil pistol, presented in a French military technical brochure published in 1899. The author is Capt. Parra.

American soldier in France in 1917; he is using a Colt 1911 pistol worn on the right in a left-hand hip holster then in service for revolvers in the cavalry. *Jean-Claude Fombaron*

Colt "Military Model" pistol. *Marc de Fromont*

The first research by John Browning concerning semiautomatic pistols dates back to 1894. He directed the research toward various mechanisms:

- gas-operated dropping gas piston (US Patent No. 580,923)
- short recoil and rotating slide (US Patent No. 580,924)
- short recoil and rotating barrel (US Patent No. 580,925)
- moving bolt (US Patent No. 580,926)

Several prototypes were to stem from this. The first system was rapidly abandoned because it was too complex and not adapted to handguns. All the prototypes developed by John Browning seem surprisingly modern, with the magazine introduced in the grip and without the complicated shapes of the Clair, Bergmann, Borchardt, Mannlicher, Mauser, and others.

From two other mechanisms, the inventor was to develop several families of automatic repeat-action handguns. He also made the corresponding ammunition, which remains unchanged to this day. His prototype for the .32-caliber (7.65 mm) pocket gun did not find any takers in the United States, so he crossed the Atlantic and proposed it to the Fabrique Nationale in Herstal, Belgium. More than a million Browning 1900 were made. He made a weapon, still with a nonfixed bolt mechanism, that was produced and marketed on both sides of the Atlantic in different versions: the Colt 1903 (7.65 mm) in the United States and the Browning 1903 (9 mm long Browning) in Belgium. The latter was also adopted by Sweden in 1907.

"Military Model" pistol with high spur hammer and the safety pin associated with the rear sight. *Marc de Fromont*

Two Browning pistol prototypes: **TOP**, the gas-operated model made in 1895; **BOTTOM**, a short-recoil-barrel experimental version used until 1900.

The short recoil mechanism of the barrel made for more-powerful pistols. First and foremost, the Colt 1900 fired a .38-caliber (9 mm) cartridge. The weapon uses a barrel that oscillates between two connecting rods: one at the front and another at the rear. The profile was already at this stage the same as the majority of pistols that have been familiar to us for more than a century: it has a rectangular, slightly sloping grip, containing the magazine, an enveloping slide containing the barrel, and the recoil spring, a frame with external hammer.

There were several variations of this model, and the US Army had already taken an interest, ordering several hundred models. Combining all variations, 4,274 Colt 1900 were produced from 1900 to 1903.

The "Sporting Model" of 1902 followed on from the "Military Model" of 1900. There are not many differences between the two weapons other than the definitive adoption of certain details present on the earlier model:

- The safety, associated with the sighting notch, disappeared.
- The slide serrations are machined at the forward of the trigger.
- The high spur hammer was replaced by a stub hammer with grid pattern.
- The grip plates are in ebonite.

There was also a "Military Model" of 1902, which had a large ring suspended at the base of the grip.

The Colt 1902 was produced from 1902 to 1929. Around 7,000 Sporting version pistols were made, and 18,000 in the Military version.

OPPOSITE: A 1902 Colt displayed with American items of the period: (cap, saber, insignia, water bottle, .38 auto cartridge). *Marc de Fromont*

Their involvement in the Spanish-America War and the Cuban expedition, along with th Philippines campaign, meant that the American were made aware that the .38-caliber revolver— which had been used in the US Navy starting i 1889 and in the US Army from 1892 onward—lacke power. In this context a return to the .45 wa necessary, which led John Browning to create new chambered pistol for large-caliber ammunition In 1905, the .45 ACP (Automatic Colt Pistol 11.43 mm caliber) made its first appearance, at th same time as a Colt for military use.

The Colt M1911 was first used on the European front in 1915, with an order of 5,000 for French tank units. *Claude Noir*

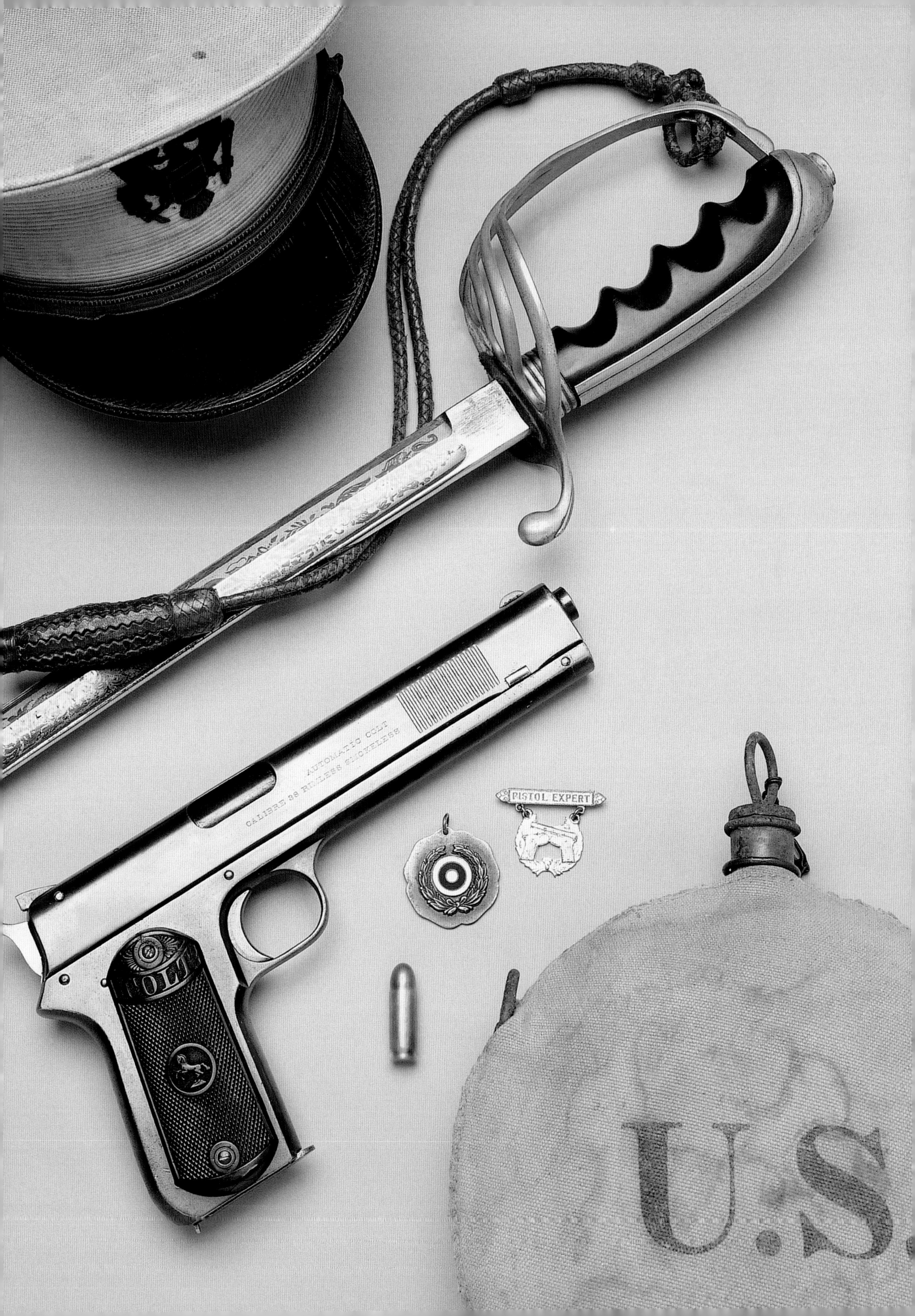
AUTOMATIC COLT
CALIBRE 38 RIMLESS SMOKELESS
PISTOL EXPERT
COLT
U.S.

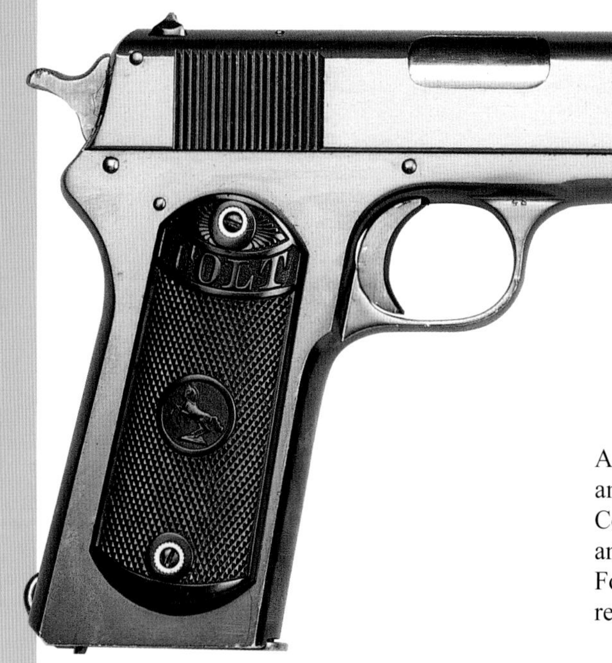

Colt 1900 with low spur hammer. *Marc de Fromont*

Compared to its predecessors, the Colt 1905 was slightly shorter and its slide was wider; the studs reinforcing both the barrel and the slide had their dimensions recalculated, and the grip plates were in walnut with a grid pattern. In total, 6,100 guns of this model were produced up until 1911.

On December 28, 1906, the US Army drew up a list of specifications that described the characteristics of the weapons likely to be presented to the board. Without entering into too much detail, it can be said that the following points were emphasized:

- **firepower:** The weapon had to use an 11.43 mm caliber cartridge and fire a bullet of at least 14.9 grams (g) at a speed equal to or more than 244 meters per second (m/s).
- **capacity:** at least six cartridges
- **safety:** allowing the transport of the loaded and armed weapon
- **simplicity:** both of its mechanism and disassembly
- **shape:** A compact shape makes for easier entry into the holster.

A number of manufacturers responded to the bid among others, the makers of the semiautomati Colt, Knoble, Lüger, Savage, Mars, White-Merril and Webley pistols but also the automatic Webley Fosberry revolver, along with the classic cylinde revolvers—a Colt and a Smith & Wesson.

Only a short time was needed for Brownin to adapt its 1905 model to the demands of th fighting forces, and the weapon that resulted wa named Model 1907; other models followed u until 1910. Tests could at last begin; the weapon were subject to trials concerning maneuverability resilience, accuracy, and use in adverse conditions The selection process was so strict that the arm decided to remain faithful to the revolver an adopt the Colt New Service in 1909.

The competition for the choice of a pisto remained open. The majority of the models teste at the beginning of the test campaign wer eliminated. Only the Colt, the Savage, and th Lüger remained in the running. Soon, the Germa manufacturer had to give way, since th modifications requested throughout the testin process were easier to apply for local manufacturers The competition therefore remained between Co and Savage.

The Colt had evolved greatly since the 190 model, thanks to the joint efforts of John Brownin and the Colt technicians. The connecting rod a the front of the barrel was removed, the gri became more sloped, and a safety lock became feature, along with an optional safety and a slid stop that could be concealed by means of a latera control lever.

Colt 1902 "Military Model."

Colt 1902 "Military Model" with safety associated with the sighting notch and high spur hammer. ***Marc de Fromont***

Another example of the use of the M1909 holster to transport the Colt M1911. ***Jean-Claude Fombaron***

As for Savage, theirs was a weapon based n US Patent 804985, filed by Maj. Searle on ·lovember 21, 1905. It was then improved by Villiam Condit. Initially the Savage Company ·as getting ready to market a 7.65 mm caliber istol, but having learned that the army had opened competition to choose a large-caliber one, the ompany hurried to develop a model in .45 caliber. 'his was ready in 1907, and two hundred of them ·ere made.

The Savage pistol operated with a delayed- pening bolt; this is a spiral movement of the arrel, with a shift of 5 degrees that ensures the elay. The weapon is organized with a recoil pring around the barrel, it comprises only forty- ve parts, and all the springs are helical.

In November 1910, new tests with the greatly eworked Colt and Savage pistols, along with a 'olt revolver, took place. Out of 600 shots fired, ie Colt New Service revolver did not experience ny firing incidents, the Colt pistol totaled twelve icidents, and the Savage totaled 43. The cavalry, ostile to the pistol, won the day, claiming that ie noise of the weapon with the bolt would ighten the horses!

With new tests on March 15, 1911, 6,000 shots were fired in series of 100, with a five-minute cooling-down period between each series and cleaning between every 1,000 shots. At the 6,000th shot, the Colt had not registered a single firing incident, while Savage was at its fortieth, with several parts broken and replaced. Browning had won!

The director of the US Army ordnance, Gen. Crozier, declared: "The Colt pistol is superior due to its safety, resistance, its ease of maintenance, and its accuracy." On March 29, 1911, the secretary of state for war approved this choice by signing the decree taking up the "Automatic Pistol, Caliber .45, Model 1911."

The Savage pistol, an unfortunate challenger to the Colt. ***Doc Savage***

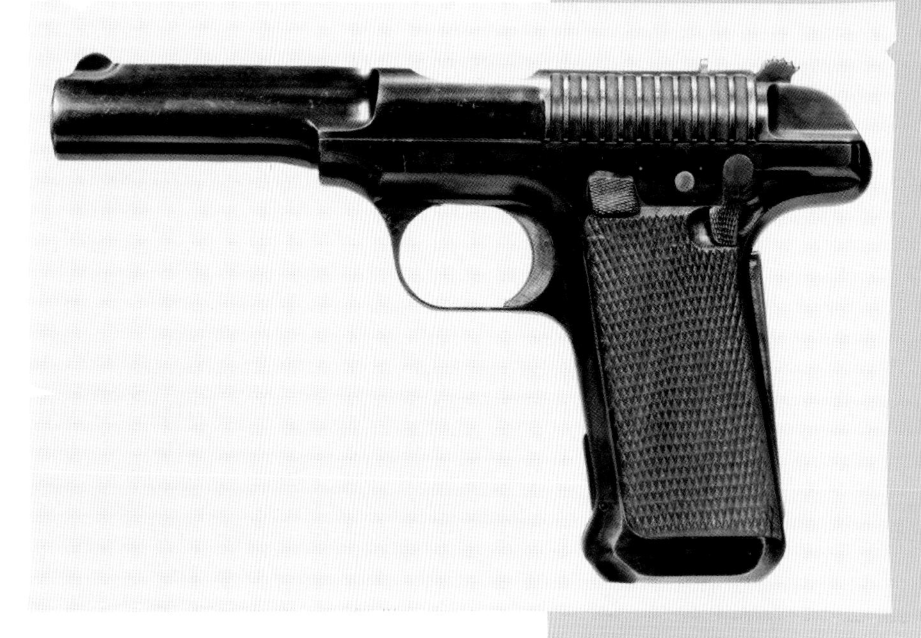

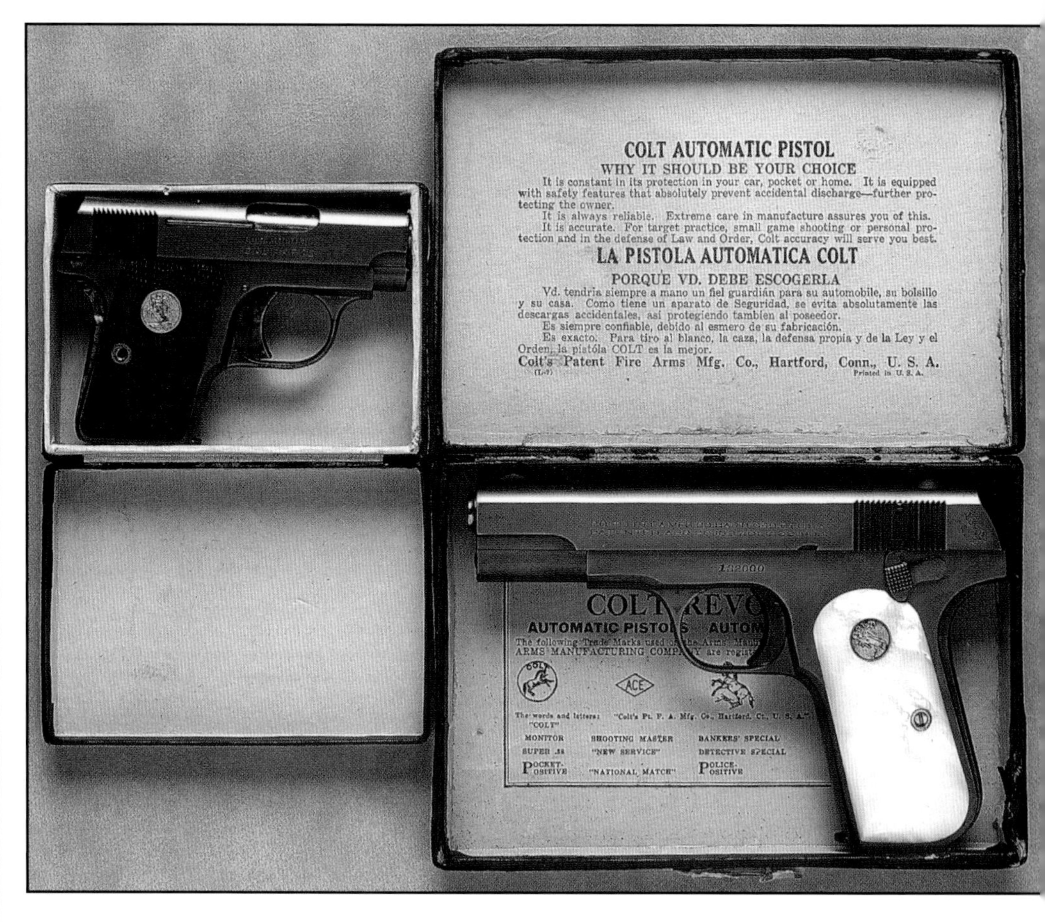

The other Colt productions: **LEFT**, a Baby 6.35 mm caliber pistol; **RIGHT**, a Colt 1903 in .380 ACP (9 mm short Browning). *Butterfield & Butterfield*

In 1913, the American armed forces were of modest number, with, in the active army, 4,744 officers and 84,810 men assigned among

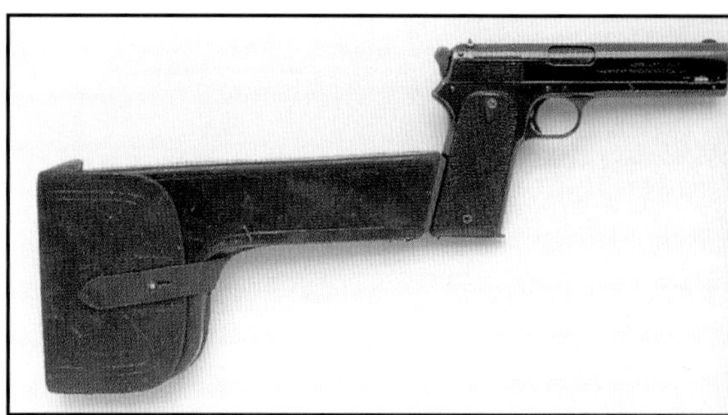

Colt 1905 pistol with stock holster. *Butterfield & Butterfield*

- thirty infantry regiments of three battalions (except the Puerto Rico regiment, which had only two),
- fifteen cavalry regiments of three squadrons,
- six field artillery regiments (three mounted, one horse drawn, and two mountain), and 170 batteries of coastal artillery, and
- three engineers battalions and twelve signal companies.

World War I was to change the course of event[s]. Starting in 1915, America received requests f[or] deliveries of foodstuffs and manufactured good[s] from the Allies (Great Britain, France, and Russia[).]

Officially neutral, the Americans responde[d] to the first requests with great care; however, the[ir] great industrial power was soon put to good us[e].

When the United States entered the war, th[e] image and the organization of their army were t[o] change. Its increase in power was swift; betwee[n] the summer and November of 1917, two millio[n] men landed on French soil.

Arrival of Gen. Pershing in France in June 1917. He and his second-in-command are carrying a Colt M1911 pistol with a two-magazine leather pouch.

Colt 1911 pistol, with two magazines. Notice the thermal treatment visible on the upper part of the magazine on the right. *Col. A. Nowak / Marc de Fromont*

HAPTER 3

THE COLT M1911

It was on December 28, 1911, that the first military Colts were assembled. The manufacturer at Hartford also delivered the weapon the civilian market starting in March 1912. The eapon was essentially the same as the military ersion but with different markings.

evelopment and Production

roduction of the Colt M1911 in the year 1912 as in the order of 17,250 total; this number was ivided fairly evenly between the US Army and e US Navy.

Between the years 1912 and 1915, minor odifications were made. These concerned the llowing:

- trigger spring
- magazine catch
- grip-plate screw: the head became thicker
- magazine: the bottom plate was modified and its lips were steel tempered
- end of the recoil spring (modified twice)
- hammer strut pin
- upper face of the sighting notch was machined flat (it was formerly rounded)
- Markings; certain innovations came into force at the end of the trial period (safety, hammer spring housing) or were brought in when it was put into service (trigger spring) since they had not yet received a patent, the date of the final patent, August 13, was added to the markings appearing on the slide.

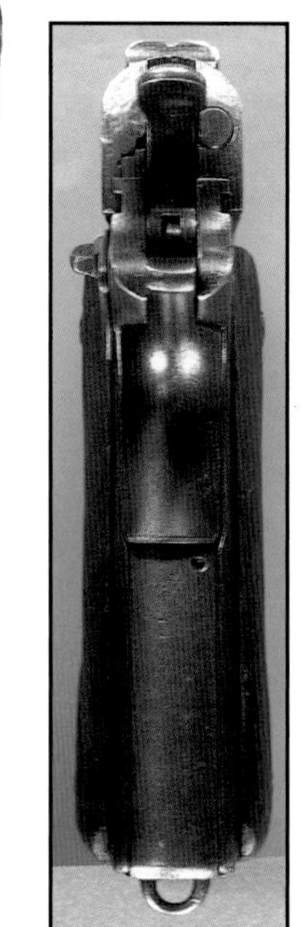

The M1911
Colt-produced pistol

The specific marking referring to the designation of the model

The back of the M1911 grip

The quality of manufacture of the M1911 pistols has always been constant. *Col. A. Nowak/ Marc de Fromont*

- the hammer, which from then on had a longer cocking piece
- a more powerful main spring to compensate for the slight increase in the weight of the hammer
- spring catch of the safety lever

But the contract between Colt and the US government stipulated that the state would have the new pistol built in its arsenals. Starting in December 1912, Springfield Armory also produced Model 1911 pistols. Apart from the markings, these were in every respect identical to those produced by the private constructor.

The first M1911 pistols saw their initial action in Mexico during the expedition against Pancho Villa.

On the eve of the United States entering the war, Colt and Springfield had produced 55,553 pistols. At the same time, Colt made more than 86,000 pistols for the commercial market.

Just under one hundred M1911 pistols made by the Springfield Arsenal were sold before 1917 in the civilian sector. They were destined for the Department of Civilian Marksmanship (DCM) and bear the marking "NRA" on the body.

MODEL OF 1911. U.S. ARMY

Things changed on April 6, 1917; was on this date that the production c pistols (along with other material) on large scale was envisaged. In May 191 the daily production was 470 weapon which went up to 1,000 in June and t 2,200 during the summer. However, th was not sufficient, and the ordnance planne to entrust other companies with th production of the M1911. The manufactu of 2,500,000 weapons was planned an entrusted to the following companies:

Canada

North American Arms Company Ltd., Quebec, which was to work with Ross,
Caron Brothers, Quebec

United States

A. J. Savage Munition Company, San Diego, California
Lanston Monotype Company, Philadelphia, Pennsylvania
Savage Arms Corporation, Utica, New York
Borough Adding Machine Company, Detroit, Michigan
National Cash Register Company, Dayton, Ohio
Remington Arms UMC Company, Bridgeport, Connecticut
Winchester Repeating Arms Company, New Haven, Connecticut

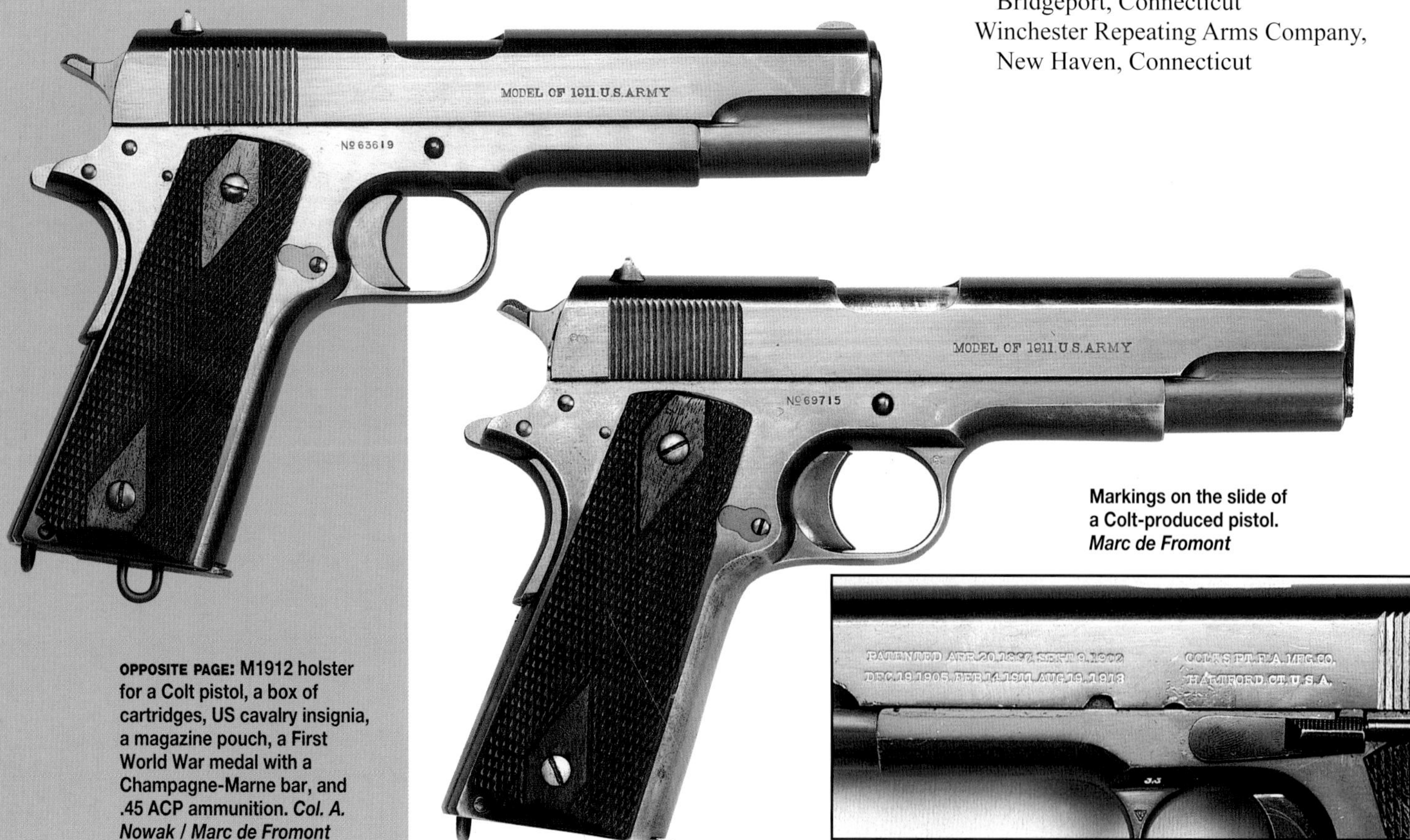

Markings on the slide of a Colt-produced pistol. *Marc de Fromont*

OPPOSITE PAGE: M1912 holster for a Colt pistol, a box of cartridges, US cavalry insignia, a magazine pouch, a First World War medal with a Champagne-Marne bar, and .45 ACP ammunition. *Col. A. Nowak / Marc de Fromont*

US
U.S.
20 PISTOL BALL CARTRIDGES, Cal. .45
Model of 1911.
For AUTOMATIC PISTOL, Cal. .45, Model of 1911.
Smokeless Powder
Muzzle velocity 800 ± 25 feet per second
Bulls Eye Powder, Lot H-48.
Manufactured at Frankford Arsenal,
Class 47, Division 1, Drawing 5.
PATENTED APR.20.1897
SEPT.9.1902. DEC.19.1905. FEB.14.1911
UNITED STATES PROPERTY
COLT'S PT.F.A.MFG.CO.
HARTFORD,CT.U.S.A.
CHAMPAGNE-MARNE

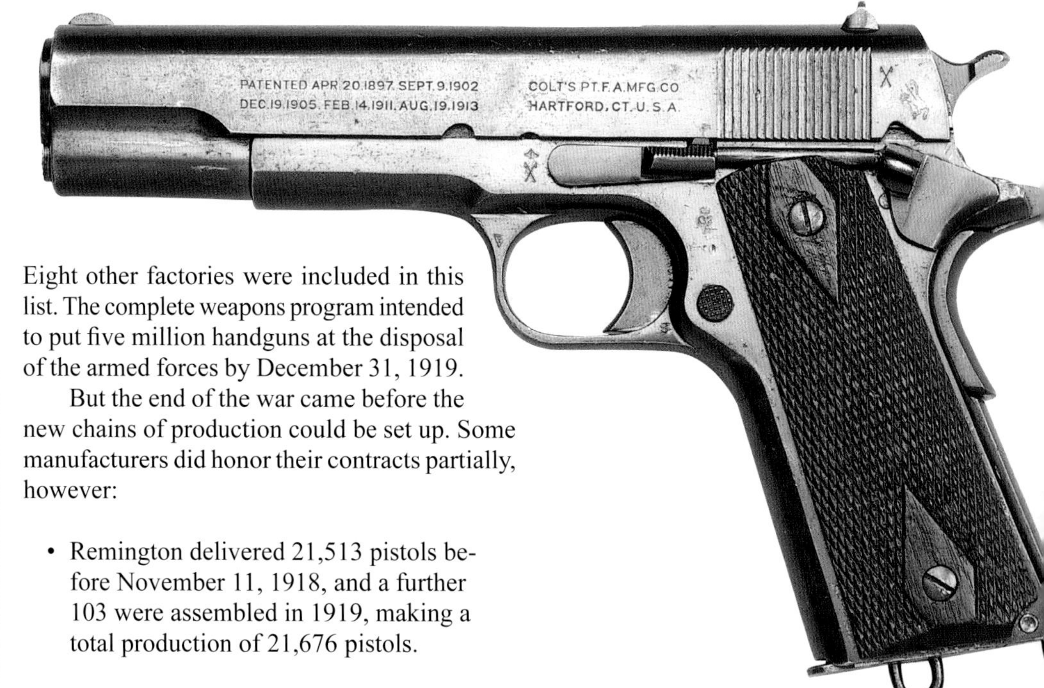

M1911 pistol intended for the British. *Col. A. Nowak / Marc de Fromont*

Eight other factories were included in this list. The complete weapons program intended to put five million handguns at the disposal of the armed forces by December 31, 1919.

But the end of the war came before the new chains of production could be set up. Some manufacturers did honor their contracts partially, however:

- Remington delivered 21,513 pistols before November 11, 1918, and a further 103 were assembled in 1919, making a total production of 21,676 pistols.
- North American Arms preproduced a hundred or so weapons.
- Savage made slides that were sent to Colt.

Close-up of the markings on Remington-produced weapons. *Marc de Fromont*

The pistols made by Remington were numbered from 1 to 21,676; however, those produced by Springfield were inserted in the Colt production. It is therefore possible to find two frames bearing the same number depending on whether they were made by Colt or Remington. The slide should have, in principle, solved the problem because of its markings, but in fact did not.

It is important to note that a good number of pistols were repaired or reassembled with parts from diverse sources. How is it possible to identify them in that case? Only the proof stamps can help the researcher, along with an important detail: the numbers on the Remington weapons are slightly larger than those on the Colt.

Other companies were entrusted with the manufacture of various parts of the pistol as well as the magazines or their components; more than six million were produced.

However, the pistols themselves were not produced in sufficient quantity, so the ordnance tasked Colt and Smith & Wesson with the production of revolvers firing .45 ACP cartridges:

- Colt made 151,700 Colt M1917 revolvers; these were weapons derived from the New Service M1909 Colt.

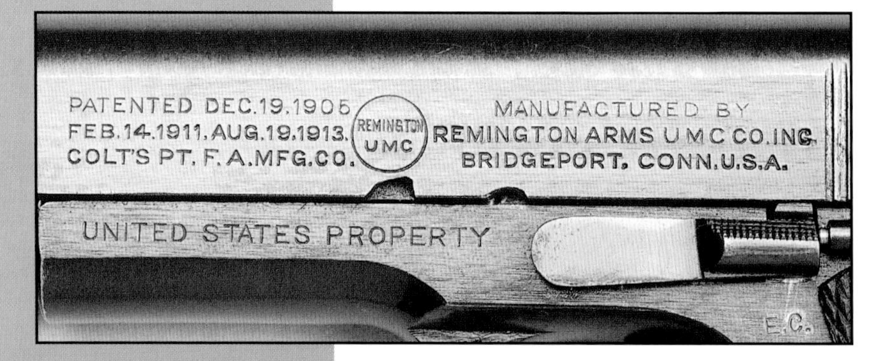

Specific markings on weapons made by Remington. *Marc de Fromont*

Right side of a Remington-produced weapon. *Marc de Fromont*

- Smith & Wesson produced 153,111 M1917 revolvers, which is an extrapolation of the New Century Smith & Wesson, which was already produced in .455 for the British.

In order for these revolvers to be able to use pistol ammunition, three were placed in a steel, half-moon clip, which limited the penetration of the cartridges in the cylinder. These cylinders themselves had been modified in order for clips to be used. After the war, a cartridge with a thick ridge was made for Colt and Smith & Wesson M1917 revolvers: the .45 Auto Rim.

At the end of the war, 529,985 M1911 pistols were delivered (487,714 of this number by Colt), at a unit price of around $15. To this, 305,011 revolvers can be added, meaning that more than half the combatants were equipped with a handgun. Out of this figure, approximately 165,000 weapons were either lost in combat, destroyed, or claimed as such (20 percent).

After the war, all the tools, equipment for checking, and unfinished parts detained by the contractors (except Colt) were retrieved and stocked by the Springfield Arsenal.

Deliveries Abroad

During the First World War, the Americans were not the only users of the Colt pistol. Norway took up the pistol in 1914 and made it under license with a modified catch control lever. This point will be dealt with later.

Russia received 51,100 pistols between February 1916 and January 1917. They were taken from Colt's commercial production, and their serial numbers ranged from C23000 to C89000.

In addition to the standard markings, on the left side of the body these pistols bore the following notation in Cyrillic script:

АНГЛ. ЗАКАЗБ

AH__.3AKA36, which is shorthand for British order, since the Allied help for the Russians was managed by the British.

Great Britain received slightly more than 17,500 M1911 pistols between May 1912 and April 1919. All were produced by Colt; 4,200 pistols were delivered in the original .45 caliber, with serial numbers preceded by the letter *C*. The others were made in .455 by Webley Auto, with a serial number starting at 10001 preceded by the letter *W*.

These British pistols necessitated a special magazine and a specific boring inside the grip.

Apart from the prefix *W* mentioned previously, British pistols bore the .455-caliber mark and proof stamps (including the "Broad Arrow" mark), along with the mark "RAF" for those allocated to the Royal Air Force (formerly the Royal Flying Corps [RFC], which was reorganized toward the end of World War I). They remained in service until 1942, and from this period onward they were transferred to the Fleet Air Arm, which assigned them to crews of rescue seaplanes.

France had already tried the Colt pistol before World War I; example no. 2144 had been sent to the government on January 31, 1913.

Wrapping of a packet of seven Webley .455 Mark I cartridges for semiautomatic pistols. In order to make it clear for the user, a "NOT FOR REVOLVERS" stamp was added. ***Marc de Fromont***

Marking on the Russian contract. ***Marc de Fromont***

.45 ACP cartridge

Base of the grip on a Smith & Wesson M1917 revolver. ***Marc de Fromont***

The Colt M 1917 revolver adapted to the .45 ACP was a serious boost to the weaponry of the American Expeditionary Force. ***Col. A. Nowak / Marc de Fromont***

The Colt .45 has remained the weapon for individual use "par excellence" over many decades (seen here in the hands of a South Vietnamese soldier).

At the beginning of the conflict, the French delegation responsible for purchasing military equipment in the United States obtained twelve pistols from Colt, which were sent on February 6 and April 26, 1915; they bear the following numbers: C14033, C15032, C15077, C15668, C15764, C16118, C16121, C16125, C16140, C16233, C16338, and C16339.

Subsequently, larger deliveries to France were registered, totaling 5,012 pistols:

- 2,000 pistols sent on May 12, 1915 (serial numbers between C17840 and C24464)
- 1,450 on December 20, 1915 (numbers between C19922 and C28371)
- 1,050 on January 24, 1916 (numbers between C22089 and C25964)

These weapons did not carry any markings other than those stamped by the manufacturer.

Later on, in December 1917, the permanent purchasing board appointed by the French government managed to obtain 500 pistols and 1,000 magazines from the Springfield Arsenal. These weapons bear military markings, but it is not known whether they were produced by Colt or Springfield, and their serial numbers are not known.

These pistols were given to tank crew; they used a brown leather belt identical to that of the Ruby pistol, but larger.

OPPOSITE PAGE: an aviator of the Royal Flying Corps (precursor to the RAF) and his Colt .455 Auto

Colt also supplied other pistols to other countries:

- **Canada:** 5,000 pistols between 1914 and 1921; the Canadians stamped their hallmark at the rear of the slide on the right side. In addition, on some models used in aviation, an RCAF marking and the squadron number is found. This marking features on the back of the grip on the hammer spring housing.

COLT AUTOMATIC
CALIBRE 455

Identification of models destined for Great Britain in 1914–18. Marc de Fromont

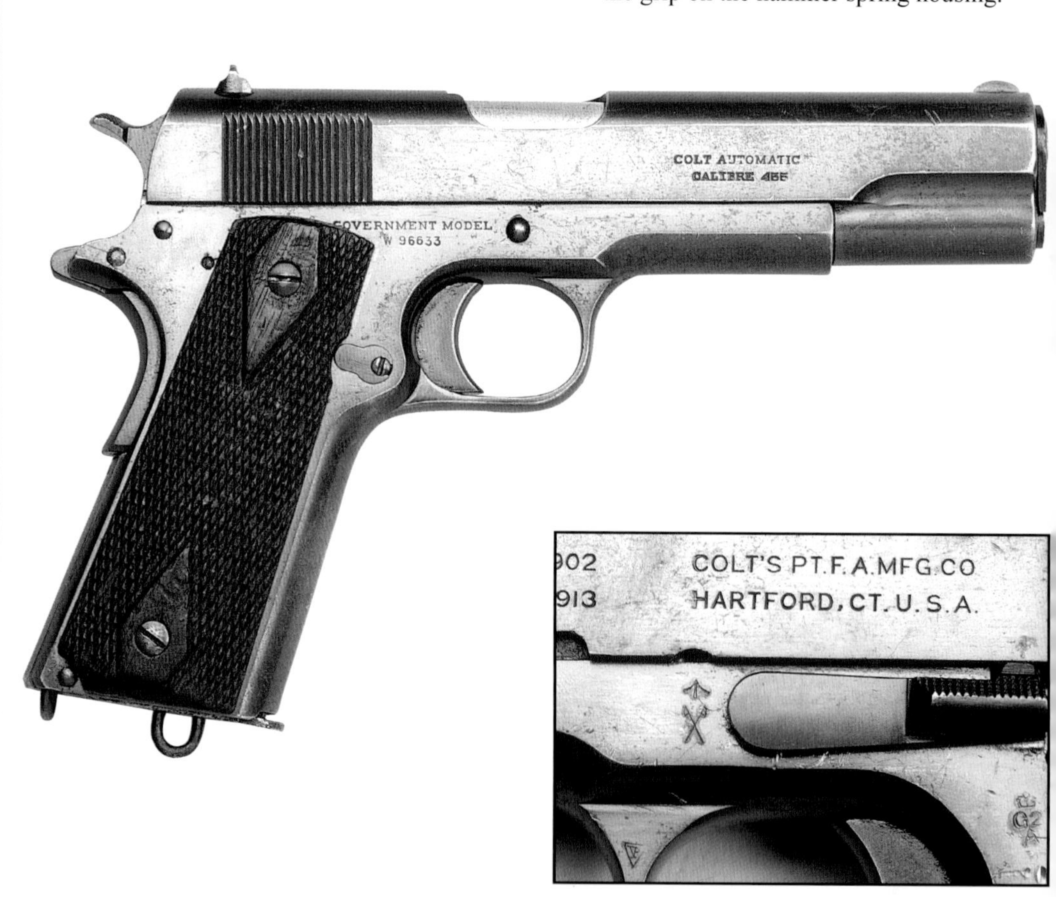

M1911 pistol chambered in .455 Webley Auto. Marc de Fromont

Weapons destined for the British army were retested from the moment they arrived! Marc de Fromont

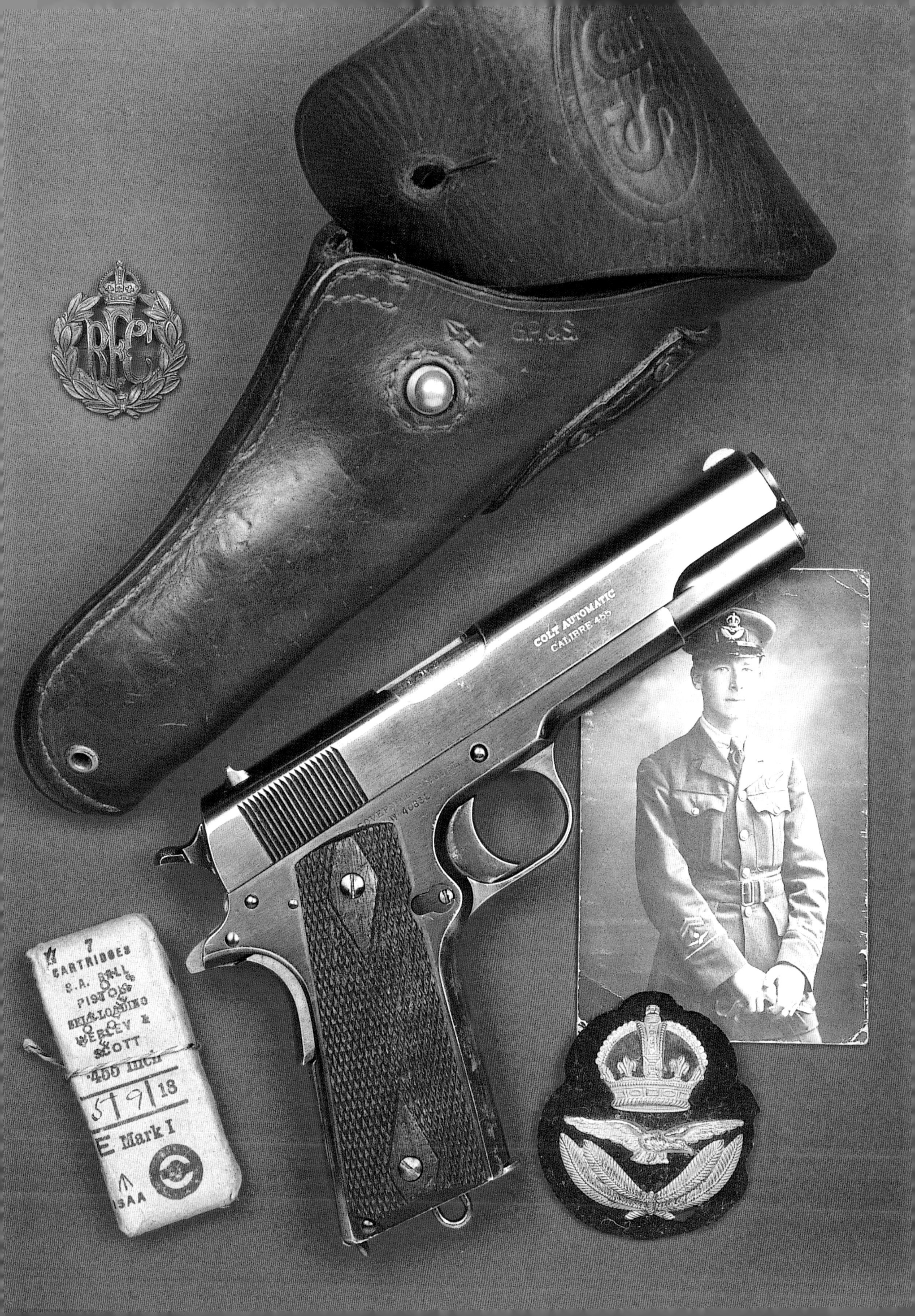
G.P.&S.
COLT AUTOMATIC
CALIBRE 455
7
CARTRIDGES
SCOTT
18
E Mark I

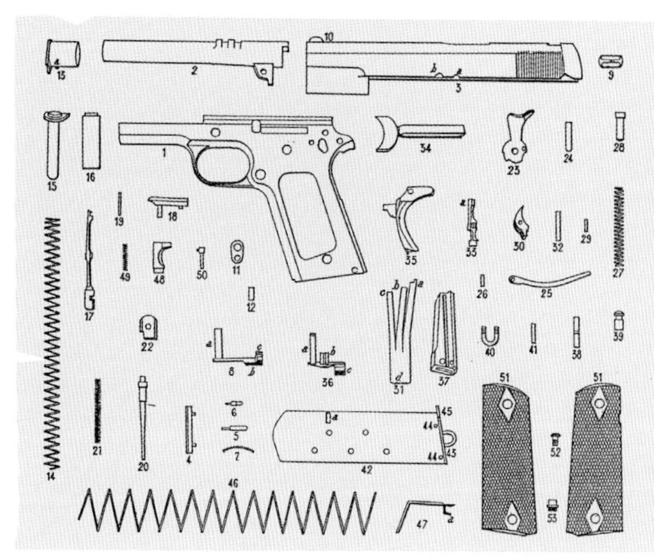

An appendix to the French instruction manual on the M1911 pistol (1915 edition)

- **Norway:** for the navy, 400 examples of the 1911 model with the slide stop normal in 1915 (nos. C18501 to C18850) and 300 others in 1917 (nos. C88901 to C89200), but with a specific marking of "11.25 m / M AUT. PISTOL M / 1914 No. XXX." on the right side of the slide, whereas on the opposite side, the year of manufacture appears. See also the production under license in chapter 9.

- **Argentina** (which adopted this pistol in 1914)**:** 321 pistols were supplied to the navy, with a specific marking of "MARINA ARGENTINA" to the right of the slide (nos. C6201 to C6400, then C11501 to C11621), and then 1,000 pistols in 1915 for the armed forces, with numbering between C20001 and C21000 and the national coat of arms on the top of the slide. Another delivery took place in May 1919. Later on, the Colt M1911A1 was adopted under the name M1927 (see chapter 9).

- **The Philippines:** 302 examples in May 1917 (nos. C86000 to C97900)

- **Mexico**

DEPARTMENT OF THE ARMY TECHNICAL MANUAL TM 9-2951-1
DEPARTMENT OF THE AIR FORCE TECHNICAL ORDER TO 11W3-3-3-42

FIELD MAINTENANCE
CAL. .45
AUTOMATIC PISTOLS
M1911 AND M1911A1

DEPARTMENTS OF THE ARMY AND THE AIR FORCE
JULY 1957

Manual of the American Armorer (1957 edition)

Description

The frame was made in steel; it supported the firing mechanism, received the barrel, and on the upper part bears the slide guideways.

The trigger and the slide are mounted in the body. The frame constitutes a fixed unit formed of the hammer, trigger, divider, and safety lock, along with their respective springs and axle pins. The hammer spring housing is positioned in a hollow at the back of the grip.

The hammer has a wide spur with a grid pattern.

The grip is fitted with walnut grip plates with a grid pattern, with a characteristic diamond shape level with the fixation screws. Bakelite grip plates of the same appearance were used by Colt at the end of the war. The base of the grip bears a ring that allows a strap to be attached.

The magazine is contained in the grip and is held in by a lever situated at the base of the trigger guard. It has seven cartridges arranged in a single stack. The bottom plate of the magazines produced by Colt was assembled with rivets, whereas those with magazines from the Springfield Armory had a fold in the lateral part of the inner surface of the magazine. The ring on the bottom plate on the original magazines was not present on war production models.

The slide is connected to the barrel by an assembly of two studs. This slide is fitted with an ejection port situated on the right. The sides bear vertical ridges at the rear for gripping.

The barrel is mounted on the body by means of a connecting rod; it has six grooves on the left, with a rate of twist of 16 inches (406 mm). The recoil spring and its buffer pitot are under the barrel, and the spring presses against the end piece held by the case, which guides the barrel at the front of the slide.

The gunsights are formed by a sighting notch fixed in a "U" added to the slide and a half-moon foresight.

The Colt M1911 is fitted with a bolt latch that immobilizes the slide is a rear position after the last cartridge is fired. Its control catch is

Markings on the slide for Colt. *Marc de Fromont*

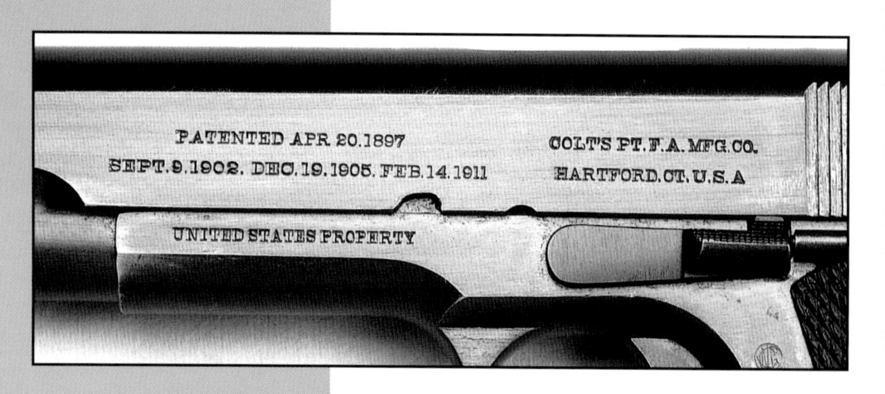

situated over the trigger. This catch is combined with the disassembly axle pin.

An optional safety lock, positioned on the hooked part at the rear of the pistol grip on the left side, enables the slide and the hammer to be immobilized when the latter is in an armed position.

The M1911 pistols, civilian or military, received a bronzed finish, but the quality of the polishing of the metallic parts could vary depending on the origin of the weapons and the period of production.

Markings of an M1911 Colt-produced pistol

Markings

The M1911 pistols bear a certain number of markings on their constituent elements. On the left of the slide are the dates of the principal patents, the business name, and the address of the manufacturer, sometimes with its logo, which for Colt was this:

PATENTED APR. 20. 1917 SEPT 9. 1902
COLT'S P.T. F.A. MFG. CORP
DEC 19 1905. FEB. 14. 1911. AUG. 19. 1913
HARTFORD. CT. U.S.A.

for Springfield Armory:

PATENTED APR. 20 1897
SPRINGFIELD ARMORY
SEPT. 9. 1902 DEC. 19. 1905. FEB. 14. 1911–U.S.A.
COLT'S PT. F.A. MFG. CO.

for Remington:

PATENTED DEC 19 1905 – MANUFACTURED BY FEB. 14. 1911. AUG. 19. 1913
REMINGTON ARMS UMC CO. INC.
COLT'S PT. F.A. MFG. CO. – BRIDGEPORT CONN. U.S.A.

for North American Arms:

MANUFACTURED BY NORTH AMERICAN ARMS CO. LIMITED
QUEBEC, CANADA

Markings of an M1911 produced by Springfield Armory

Pistols destined for the US Army bear the words "United States Property." *Marc de Fromont*

The serial numbers on pistols destined for the British are preceded by a W. *Marc de Fromont*

with these variations for the first deliveries:

MODEL OF 1911 U.S. NAVY

or, more rarely,

MODEL OF 1911 U.S.M.C.

Commercial weapons are marked as such:

COLT AUTOMATIC
CALIBRE .45

On the left of the body, on the weapons destined for American armed forces:

UNITED STATES PROPERTY.

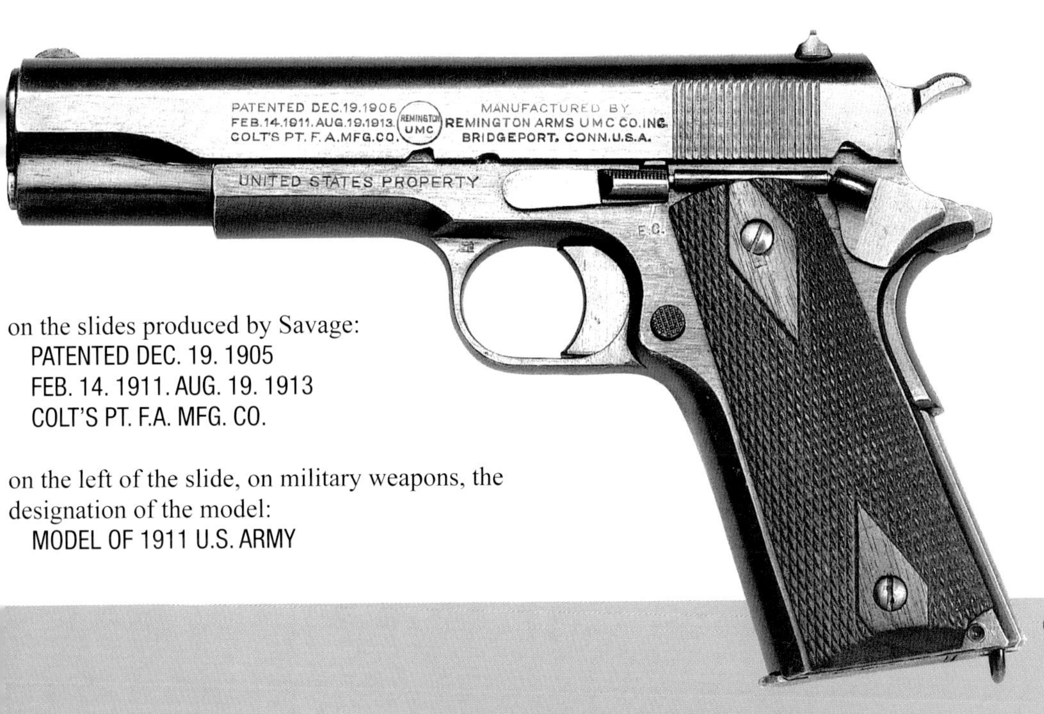

An M1911 Remington-produced pistol

on the slides produced by Savage:

PATENTED DEC. 19. 1905
FEB. 14. 1911. AUG. 19. 1913
COLT'S PT. F.A. MFG. CO.

on the left of the slide, on military weapons, the designation of the model:

MODEL OF 1911 U.S. ARMY

The military inspection stamp is stamped on the left side of the body, behind the trigger guard. *Col. A. Nowak / Marc de Fromont*

Different magazine bases. *Colin Doane*

On the right of the body, the serial number. Commercial weapons received

- the marking GOVERNMENT MODEL and
- the serial number preceded by the letter C (apart from the exceptions noted earlier).

Inspection Stamps and Other Markings

All the M1911 pistols destined for the American armed forces bear military inspection stamps; their initials are generally together within a circle.

These stamps appear on the body of the weapon, on the left side above the magazine catch. The inspectors operating before the First World War were

- Lt. Col. Gilbert H. Steward (Colt),
- Maj. Walther G. Penfield (Colt and Springfield),
- Lt. Col. John M. Gibert (Colt and Springfield), and
- Maj. Walther T. Gordon (Colt).

Other inspectors used a stamp with a simpler design, with the initials appearing on a single line:

- L.E.B. for Capt. Leroy E. Briggs (Remington)
- E.E.C. for Maj. Edmund E. Chapman (Remington)

In addition, there were preinspection and manufacturer stamps on certain parts:

- the grenade of the ordnance on the body and the slide

OPPOSITE PAGE: The Colt M1917 revolver and the M1911 pistol are two inseparable weapons of the First World War. *Col. A. Nowak / Marc de Fromont*

There are various models of magazines for the M1911 pistol, showing significant details of construction and finish, but they all are interchangeable. *Marc de Fromont*

MODEL OF 1911 U.S. ARMY
20 PISTOL BALL CARTRIDGES, Cal. 45
Model of 1911.
For AUTOMATIC PISTOL, Cal. 45, Model of 1911.
Dupont Powder Lot No.

Variations of magazines; note the one on the far right marked CAL. 455 ELEY. *Marc de Fromont*

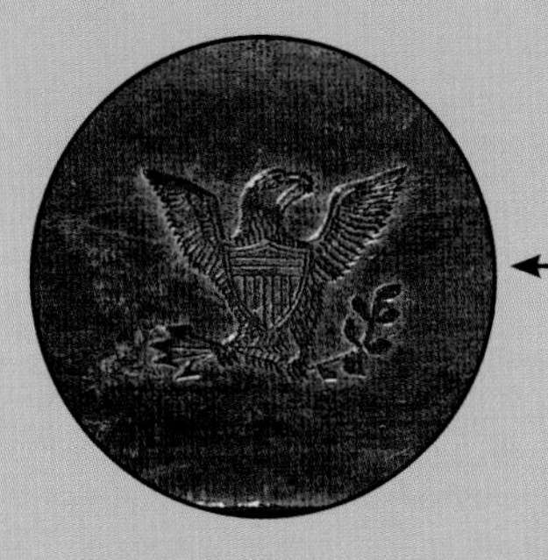

Markings and logo particular to the weapons built at the Springfield Arsenal. *Colin Doane*

The M1911A1 pistol was soon to replace the models described in this chapter. *Colin Doane*

- an eagle's-head stamp accompanied with an alphanumerical code on the left side of the trigger guard
- *H* on the slides
- *H* or *HP* or *PH* on the barrel made by Colt; this marking can also figure on the rear side as well on the top of the barrel.
- The barrels made at the Springfield Armory are marked with the letters *M*, *MD*, *J*, or *S* on the side or the letters *S* and *P* or *SP* (or both) on the mainspring.
- *P* on the mainspring or the top of the barrels produced by Remington
- a small *S* on the secondary parts of the weapons produced by Springfield
- Some identification letters are found on the bottom plate of magazines produced by contractors.
- *A*: American Pin Company, Waterbury, Connecticut
- *B*: Barnes & Kobert Manufacturing Company, New Britain, Connecticut
- "L.M.S.": Little Manufacturing Company, Hartford, Connecticut
- *R* near the magazine top: Ridson Tool & Machine Company, Naugatuck, Connecticut
- *R* near the rear: Raymond Engineering Company, New York, New York

The magazines produced by Colt, Springfield, and Remington do not bear any particular mark.

Logos

On the weapons produced by Colt, the company emblem of a colt is engraved on the slide on the left side:

- on the first models, at the rear next to the slide grips, initially with a circle and later without the circle
- then, between the dates of the patents and the business name

The pistols made by the Springfield Arsenal bear an American eagle on the right side of the slide.

Those produced by Remington are struck with the brand logo on the left slide, between the dates of the patent and the business name of the manufacturer.

Lt. Thorby of Chicago, of the 41st A.I.R. Seen here, in Canicatti, Sicily, on July 14, 1943, blasting the lock on a house with his Colt .45; a sniper was suspected of seeking refuge inside. *US National Archives*

Colt-made M1911A1 pistol

CHAPTER 4

THE COLT M1911A1

During the First World War, several minor flaws of the Colt M1911 were pointed out, but the period was not one of excessive perfectionism. At the end of the conflict, all current contracts were canceled, although the needs of the American armed forces in peace time were considered insufficient at January 14, 1919.

Development and First Orders

The number of weapons available for service (pistols and revolvers together) was 669,885, and the requirements were estimated at 1,100,544. A large difference, but the supply of additional weapons did not take place until 1924.

Attempts were made to improve the pistol during the period of respite. A well-known marksman, Marcellus Rambo, suggested certain innovations that the Cavalry Inspection then went on to test. The observations were made known both by the cavalry and infantry.

The main criticism made concerning the M1911 pistol involved the hammer. From the moment the cocking piece was lengthened, arming was easier, but during the rear movement the cocking piece had the tendency to pinch the skin between the thumb and the index finger on those users with larger hands. The solution consisted of holding the pistol higher up on the hand, but then it became difficult to squeeze the trigger, and accuracy suffered as a result.

A certain number of suggestions that resulted from this were transmitted to Colt: the size of some parts (frame, slide, foresight, recoil spring) was recalculated, while others had their profile slightly modified (safety catch, frame latch, sighting notch, butt plate). The caliber of the barrel on the hill of the rifling was taken back from 11.3 to 11.25 mm, and the depth went from 0.076 to 0.089 mm.

In 1924, Colt delivered 10,000 pistols modified in this way; they are numbered from 700001 to 710000.

These pistols, called "transition" (of which a number were designated to become the National Match), were subject to a campaign of experiments led by the Infantry and Aviation Inspection and then by the Ordnance Corps board, which retained the major part of suggested improvements on the condition that the interchangeability of parts with weapons already in service remained possible.

Marking on an M1911A1 produced by Remington

Diagram of the parts of an M1911A1

OPPOSITE PAGE: M1911A1 pistol, on a camouflage, of a model almost exclusively used in the Pacific, with its M 1916 holster, compass, and magazine pouch. *Col. A. Nowak / Marc de Fromont*

On May 17, 1926, the improved pistol was adopted under the name "PISTOL AUTOMATIC CALIBER .45 M1911A1."

However, the country was in the middle of an economic depression; it was impossible to envisage new orders of pistols, and this continued up until 1937. In the meantime, it became evident that complete interchangeability of the M1911 parts was not total, depending on whether one was dealing with a Colt-, Springfield-, or Remington-produced weapon. The ordnance therefore examined all the inspection tools, revised all the plans, and fixed a total resizing of all constituent elements in the event that production resumed. The operation was completed on January 15, 1936.

The following year, Colt received an order for 1,580 pistols. This was designed both to satisfy the needs of the navy and to renew stocks at the Springfield Arsenal. This order was in three lots, in which final finishes were made at the final hour.

Control stamp on the barrel

During the First World War, more than half the soldiers declared they had lost their weapon in combat, and these losses are estimated at around 165,000. *Jean-Claude Fombaron*

- On the first lot (numbers 710001 to 711000), the slides are still marked "MODEL OF 1911 U.S. ARMY," since (almost) everything was thought of except the principal marking! The occasion was seized to enlarge the notch on the divider at the base of the slide and to add a muzzle cap in hard steel at the level of the firing-pin hole. Weapons received a bronzed finish on a prebrushed surface.

- With the second lot (numbers 711001 to 711605), new markings for the proof stamps were adopted; in particular, a *P* was stamped on the left side of the barrel base as well as on the left side of the body, above the magazine catch.

- The weapons of the third lot (numbers 711606 to 712349) are identical to the previous lot, and they left the factory in June.

Of those numbered from 710001 to 711605, 836 were delivered to the navy, and 769 to the army and of those numbered from 711606 to 712349 744 were delivered to the navy.

US
MODICA.
B.J.
849798 C
TYPE-AB
T.7/43
USMCR
M 1911 A1 U.S. ARMY
UNITED STATES PROPERTY
№ 1218796

Throughout the following years, the Colt Company received other orders designed to satisfy the needs both of the army and the navy:

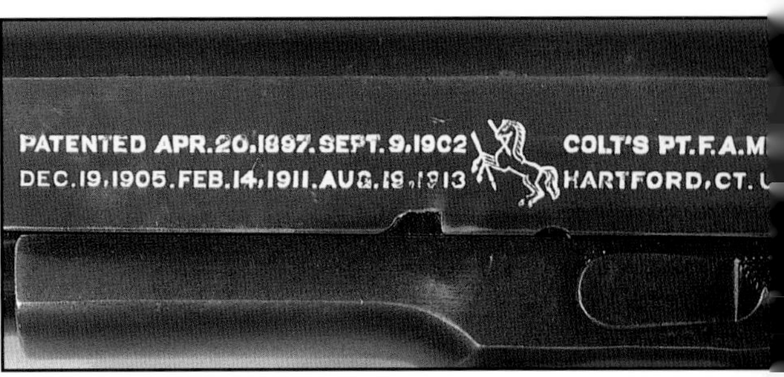

Marking on a Colt-produced slide

- 1,296 pistols in 1938 (nos. 712350 to 713345)
- 3,636 pistols in 1939 (nos. 713646 to 717281); the cocking piece is slightly resized and made shorter. These weapons were delivered to the US Navy.
- 4,696 models in 1940 (nos. 717282 to 721977); plates in plastic were tried to replace those in wood (plates in aluminum and bakelite had been previously tested).

Marking on a body produced by Colt

In 1941, Colt saw its monthly production grow, and it tripled between the beginning and the end of the year. The factory at Hartford supplied 34,756 pistols (nos. 721978 to 756733); the plates in plastic became the norm, and Parkerizing was replaced by a bronzed surface treatment.

Prewar Production

Starting in 1938, the US defense secretary put into place a vast plan with the objective of ensuring the conversion of industrial sites to weapons production sites and of adapting equipment in service to be able to cope with mass production.

It was in this context that the study of the mass production of the M1911A1 pistol was entrusted to the Singer company. This firm specialized in the manufacture of sewing machines and had a solid experience in mechanics. The Singer plan was then submitted to a certain number of weapons enterprises, which were asked to perform feasibility studies on various operations at the same time as taking into account the criteria of price, quality, and interchangeability.

In light of the results, the American government, which held the license for the production of the Colt pistol for its own needs, ordered substantial quantities of M1911A1 pistols from the end of 1940 onward. Other production was launched—with the United States entering the war at the end of 1941—with the participation of a multitude of subcontractors.

As for Colt, even though the firm had never given up the production of handguns, it had essentially focused its efforts on the production of machine guns of all calibers.

Prototype of a Colt 1911A1 from Singer, serial number X00 in its box. *Butterfield & Butterfield*

Colt

The company at Hartford stopped its production of commercial weapons in 1942 with no. C215083. It then concentrated its efforts on military models until the end of the conflict by supplying

- 99,367 pistols in 1942 (nos. 756734 to 856100), the weapons numbered from 793658 to 797639 were supplied to the US Navy.
- 277,048 in 1943 (one part nos. 856101 to 958100 and the other part nos. 1088726 to 1208673); a part of these pistols were taken from a stock of 6,575 unsold commercial models that were renumbered and treated with phosphate.
- 134,318 in 1944 (nos. 1609529 to 1743846) and
- around 135,810 pistols in 1945 (nos. 2244804 to 2368781; the last number was assigned to the very last pistol mounted in the framework of the military contract, even though the last number assigned to Colt was 2380013).

Thus, the total number is estimated at around 590,443 examples.

There may be differences between the figures quoted here and those mentioned in other works. We have referred to the most-recent research on the subject.

The differences that can be encountered, depending on the sources used, come from the fact that some manufacturers communicate the figure of the weapons ordered during period n, but it is necessary to take into account the fact that a part of the order was delivered during period $n + 1$, which results in a certain confusion.

Singer

The Singer Manufacturing Company, famous for its sewing machines, owned several factories and had a worldwide distribution network. It developed the famous industrial plan used during the Second World War for the production of the M1911A1.

On April 17, 1940, it received an order for 500 pistols, which were delivered at the beginning of 1942. These pistols bore a serial number between S800001 and S800500 and were given a bronzed matte finish. This order was followed by another for 15,000 parts, but the Singer Company declined this order, preferring to make material that was more compatible with its production capabilities (especially the firing-control systems for artillery). The equipment held by Singer was then transferred to other manufacturers.

Harrington & Richardson

Considered at the same time as Singer, the Harrington & Richardson firm of Worcester, Massachusetts, received an order for 500 M1911A1 pistols on April 23, 1940. Even though the deadline for the delivery was extended twice, the company showed itself to be incapable of carrying out the order, and it was canceled on June 24, 1942.

It is estimated that ten to fifteen pistols at most were mounted at H&R, since the company did not satisfy the specifications required. The numbers that had been assigned to the manufacturer were from H800501 to H8001000.

Ithaca

Ithaca Guns Company was founded in 1880; it had absorbed several armories and in 1936 became one of the leaders in armaments in the United States. This enterprise was approached in 1940 to produce M1911A1 pistols, but following technical and administrative problems, manufacture only really started in 1943.

Certain weapons were mounted with M1911 preform body casts that had been stocked at Springfield starting in 1919.

Ithaca production can be summarized as follows:

- 161,000 pistols in 1943 (nos. 656405 to 916404, then nos. 1208674 to 1279673 and nos. 1441431 to 1471430)
- 73,862 examples in 1944 (nos. 1816642 to 1890503)
- last, 100,605 in 1945 (nos. 2075104 to 2134403 and 2619014 to 2693613, even though the last number assigned to Ithaca was 2693613)

Without being formally discredited, American inspectors considered Ithaca production to be at the limit of regulation specifications. That is perhaps why the majority of M1911A1 pistols produced by Ithaca were delivered to Great Britain and other allied troops.

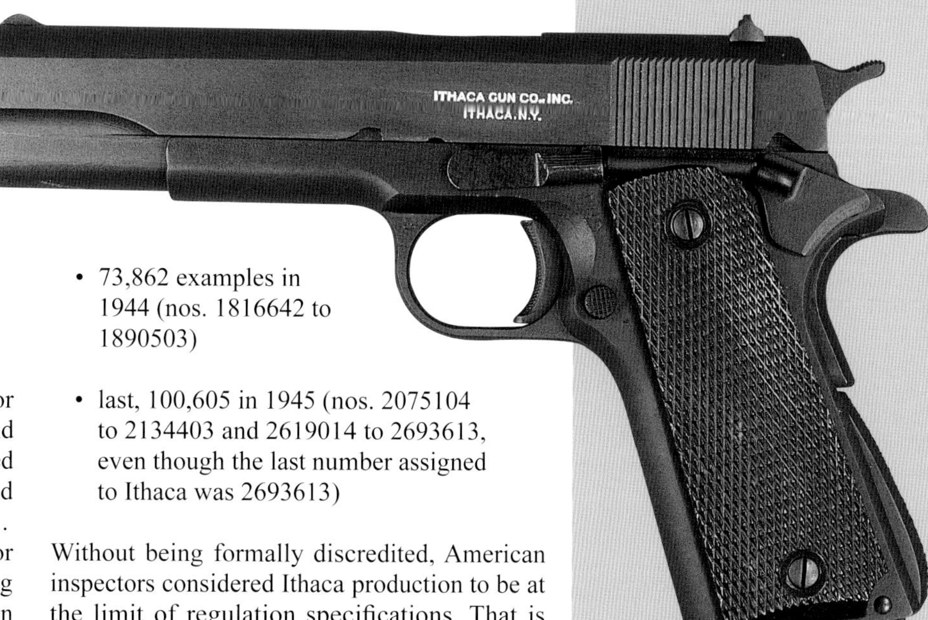

M 1911 pistol made by Ithaca

Remington

The Remington Rand Company, incorporated in the Remington group, was specialized in the production of typewriters, office equipment, and electric razors. Approached starting in 1942 to make magazines for Colt pistols, it then received an order for the delivery of complete 1911A1 models.

Remington Rand is the principal manufacturer of regulation military pistols, implementing production units in several factories of the group through using male shift workers, female labor, and even apprentices. A certain number of barrels were subcontracted to Colt, Harrington & Richardson (marking H&R on the upper stud), and the Springfield Arsenal; it is the same for certain other parts.

By the end of the war, Remington Rand had made 1,032,000 pistols:

- 424,830 in 1943 (nos. 916405 to 1041404, then 1279699 to 1441430 and 1471431 and 1809528)
- 257,395 in 1944 (nos. 1743847 to 1816641 and 2134404 to 2075103)
- 195,526 in 1945 (nos. 2134404 to 2244803 and 2380014 to 2368781, even though the last number assigned to Remington was 2619013)

Markings on the body on a Remington-produced weapon

Marking on a Remington slide

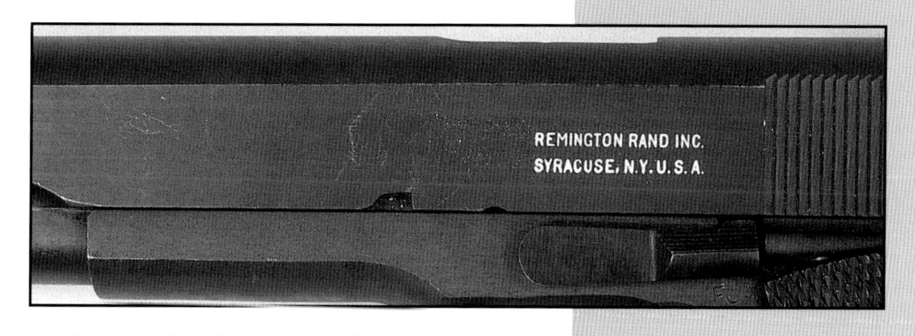

Marking on an M1911A1 produced by Union Switch & Signal

From this, the Remington Rand Company also produced the cheapest M1911A1 pistol, which had a unit price of $10 at the end of the war. To obtain this result, the company set up new methods of manufacture for certain parts without calling into question the interchangeability.

Remington Rand also produced a hundred or so pistols for its internal use, designed to test the new methods of manufacture. These weapons are numbered starting with ERRS 1 (Experimental Remington Rand Series).

Union Switch & Signal

The Union Switch and Signal Corporation (US&S) of Swissvale, Pennsylvania, is a subsidiary of Westinghouse, whose first vocation was the manufacture of signaling equipment for trains. During the first world conflict, it participated in the war effort, and it was therefore quite natural that the company was approached by the ordnance in 1942.

OPPOSITE: Two M1911A1 pistols and their accessories. *Col. A. Nowak / Marc de Fromont*

After the preproduction of a hundred pistols specially numbered EXP no. 1 to EXP no. 100, mass production was able to start at the beginning of 1943. The initial contract planned for the manufacture of 200,000 pistols, but taking into account the state of military needs at the time of the first deliveries, the contract was canceled and US&S Co. delivered a single lot of 55,500 pistols numbered from 1041405 to 1096404.

As compensation, the enterprise was tasked with the machining of casts for parts of the US M1 carbine and other military equipment.

The majority of the weapons were bronzed on a sanded surface, giving them an appearance similar to that of Parkerizing applied by other manufacturers.

Deliveries Abroad

Colt delivered 10,000 commercial M1911A1 pistols to Argentina from 1927 onward. These pistols had the following marking on the left side of the slide:

COLT'S PT. F.A. MFG. CO.
HARTFORD, CT. U.S.A.
PAT'D APR 20. 1897. SEPT 2 1902. DEC 19
1905. FEB 14. 1911 AUG 19.1913

This was accompanied by the Colt logo at the end of the two lines.

On the right side of the same slide, the armories of the Republic of Argentina had the following:

EJÉRCITO ARGENTINO
COLT .45 MOD. 1927

In addition, specific Argentine inspection stamps are found (*R*, *RA*, or *VP*). Later, this model was produced under license in Argentina with a particular marking (see chapter 9).

Moreover, the same Colt company delivered to Mexico, for an indeterminate period, the same model of pistols, bearing a marking identical to the Colt horse as well as "COLT Automatic / Caliber .45 / EJÉRCITO MEXICANO" on the right side, with the Colt horse symbol to the right of the marking.

In November 1939, Great Britain sent a buying commission to the United States with the aim of doing research on and purchasing a certain amount of military equipment. In the area of light weapons, the emissaries of His Majesty's government were to acquire US M1917 and Springfield M1903 rifles, machine guns, Thompson M1928 submachine guns, and also handguns.

If the rifles and machine guns, whose caliber was not standard with those weapons of the active armed forces, were sent to the Home Guard, the submachine guns went on to equip elite units and in particular the commandos. Handguns had various destinations. Colt supplied 14,000 pistols and revolvers of diverse models to the British, which were exported to Great

Corporal Technician William R. White, originally from Perryville, Arizona, crating M1911A1 pistols in June 1944. White belonged to the 190th Ordnance Company, located in Leda in the province of Assam, India. *US Signal Corps*

SNL-B6
QUANTITY
MAGAZINE
DRG. C-869
ITEM STOCK
ITHACA
CARTRIDGES
DUMM
CALIBER .4
UNITION LOT E. C.
EVANSVILLE ORDNANCE PLANT
UNITED STATES PROPERTY
M 1911 A1 U.S. ARMY
NO. 991940
U.S.
CARTRIDGES,
115
REMING
BRIDGEP
FOR USE IN
EFFECTIV

The Allies were copiously supplied by the Americans: here, a flag of the 1st French Foreign Legion Cavalry Regiment, with its guard of honor presenting arms with Colt M1911A1 weapons. *Képi Blanc*

Markings on a British Colt ex-pistol. *Colin Doane*

Britain by Winchester or other nongovernmental organizations (officially, the United States was neutral!). Among this number were 1,441 .45-caliber pistols, the majority of which were the 1911A1 model in a bronzed finish, as well as several others in a nickel finish or National Match type. Colt also supplied 1,429 pistols of the same type in .38 Super Auto.

Moreover, thousands of individuals gave their own weapons to the British, to help them repel the invader. These weapons were grouped together in shopping centers, armorers, or police centers. A wide variety of weapons were found among them, even Colt .45s from different sources.

In March 1941, the United States set up the famous Lend-Lease Act, which enabled the US to strengthen their defenses and sell equipment, weapons, and food supplies to nations at war, which would be paid for later once peace had been reestablished. In this manner they could ensure both the prosperity of local industry and the freedom of oppressed peoples. In short, apart from the Axis powers, everybody was satisfied.

TECHNICAL CHARACTERISTICS	
Type of weapon	Semiautomatic pistol
Model	1911 and 1911A1
Nationality	American
Caliber	11.43 mm
Ammunition	.45 ACP
Percussion	Central
Function	Short barrel recoil
Trigger	Single action
Feeding	Box magazine
Total length	220 mm
Barrel length	128 mm
Height	137 mm
Thickness	33 mm
Barrel grooves	6 on left
Empty-weapon weight	1,180 g
Magazine capacity	7 shots
Trigger pull	2.5 to 3 kgp

In this context, thousands of rifles, carbines, submachine guns, machine guns, rocket launchers, tanks, artillery parts, planes, and ships, and millions of cartridges and shells, were sent to the Allies.

In addition to the weapons mentioned previously, 78,625 M1911 and M1911A1 pistols were delivered to the British.

Despite the critical situation, administrative rules continued to be observed. Also, the majority of delivered weapons were retested and marked with a superb stamp, "NOT ENGLISH MADE," struck on the right side of the body. This marking was accompanied by diverse stamps (those of London and Birmingham in particular), as well as the inevitable Broad Arrow, the mark of the ordnance.

To round the process off, when these weapons were withdrawn from service they received a new (farewell) stamp: RELEASE BRITISH GOVT. 1952.

At the same time, the Canadians received 1,515 pistols and the Free French forces got between 50,000 and 60,000 weapons of the same type, while 15,692 .45-caliber pistols were parachuted to the French Resistance by the UK Special Operations Executive (SOE) and the US Office of Strategic Services (OSS).

The resistance organizations in Europe additionally received 8,000 to 10,000 Ballester Molina pistols produced in Argentina.

Some M1911/1911A1 English, Canadian, or American pistols were retrieved by the Germans and were used under the name of Pistole 660(a).

Description

Generally the Colt M1911A1 pistol is organized like the M1911, but there are a certain number of modifications, along with more-technical differences, making the two models distinct:

- Curved hammer spring housing at the rear of the grip; its surface is striated in a grid pattern.
- Shorter trigger; its grip surface is striated.
- The rear angles of the trigger guard are chamfered.
- The foresight is wider.
- The shorter hammer has a narrow cocking piece.
- The grip plates are in a brown-plastic material with grid pattern surfaces.
- The weapon has a phosphate, or sometimes sanded or bronzed, finish.

These are the general differences, but each one of the contractors and subcontractors made a multitude of "subvariants" for a number of elements:

- **Ring strap:** a part produced by Colt, Humasin Mfg. Co.
- **Bolt catch:** a part made by Colt, Ithaca, and Remington Rand
- **Crossbolt, sleeve pin, and screws:** parts made by Colt, Hartford Screw Machine Co. (in Hartford, Connecticut), Lux Clock Co. (in Waterbury,

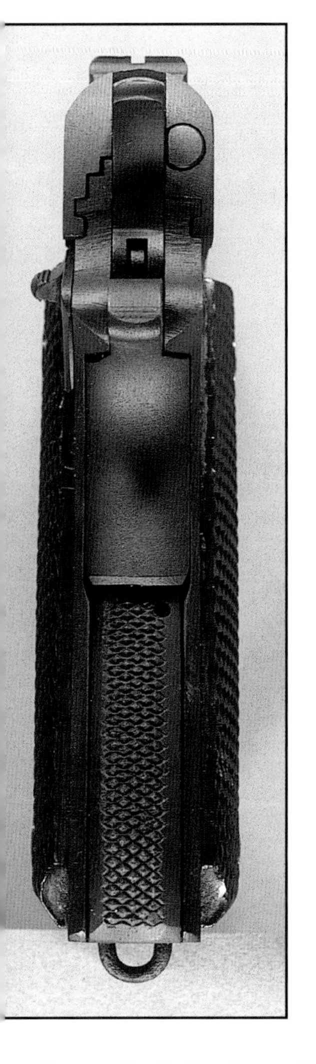

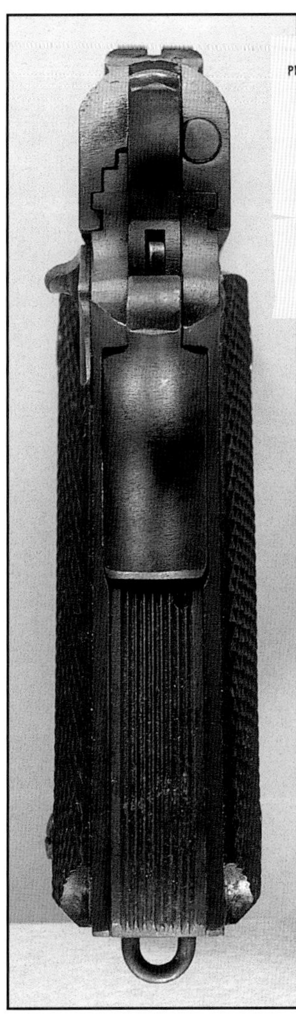

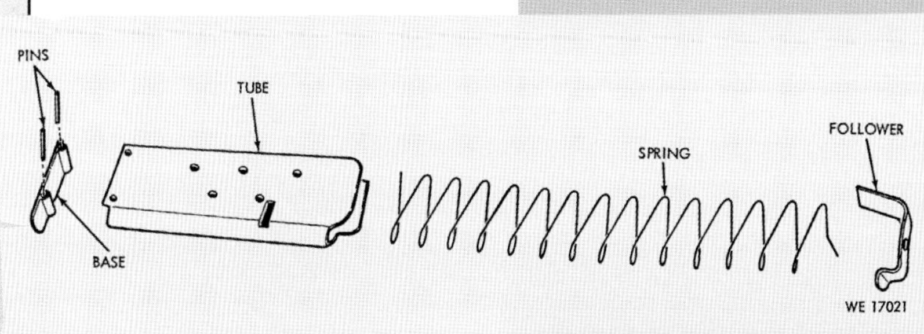

Constituent elements of a magazine

Variations in the manufacture of hammer spring housings

- **Firing-pin cap:** made by Colt, Remington Rand Elmira, and Remington Rand Ilion

- **Barrels:** not all barrels were produced by contractors; some of them were made by Colt, Springfield Armory, High Standard Mfg. Co., and Flannery Bolt Co. for the whole of the production. The boring of the barrel's flat locking surfaces was 11.30 mm on the M1911, 11.25 mm on the arms supplied in 1924, and 11.22 mm on the M1911A1.

- **Body:** the bodies are identical on all models; they bear the serial number and inspection stamps.

Connecticut), Remington Rand Elmira, and Remington Rand Ilion (both subsidiaries of Remington Rand Syracuse)

- **Connecting rod:** they have various origins, including Atlantic Service, Botnick Motors (in Birmingham, New York), Colt, Remington Rand Elmira, and Remington Rand Ilion.

- **Magazines:** the magazines produced by Colt are not marked, but a lot of others bear the markings of the subcontractors at their base. These include the following companies:
  - General Shaver Division of Remington Rand (*G* marking); this firm produced magazines with two types of magazine follower, with a smooth follower and a ribbed one.

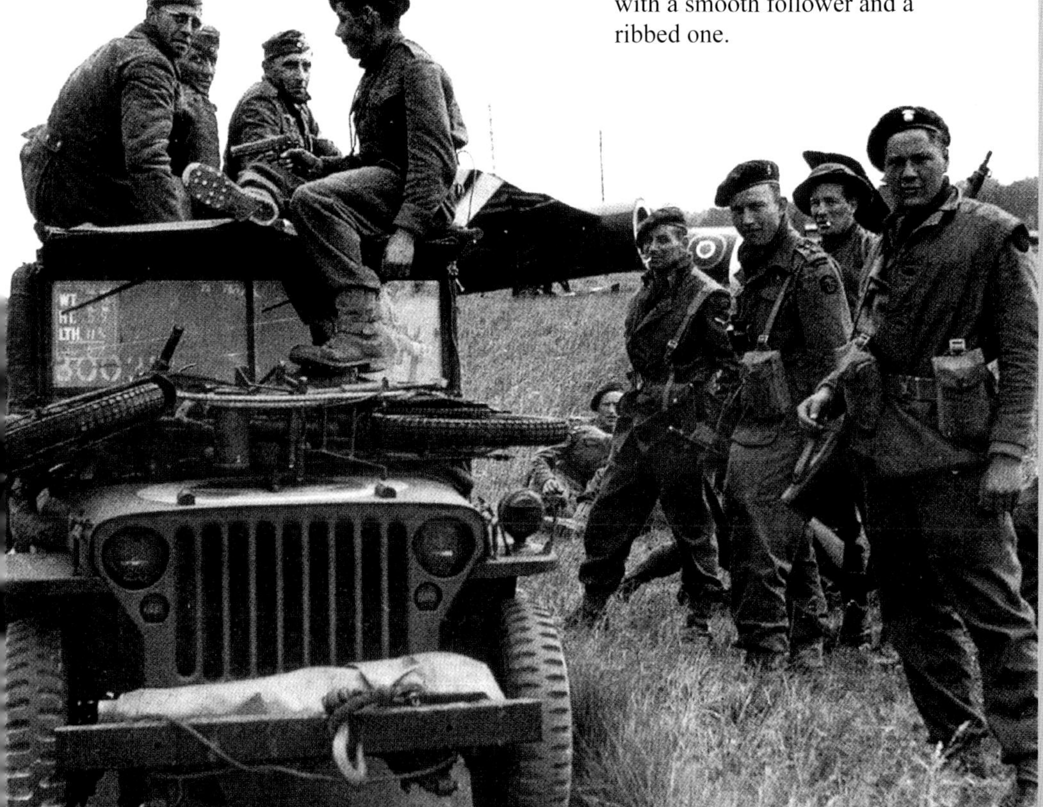

British airborne troops in Normandy on June 6, 1944

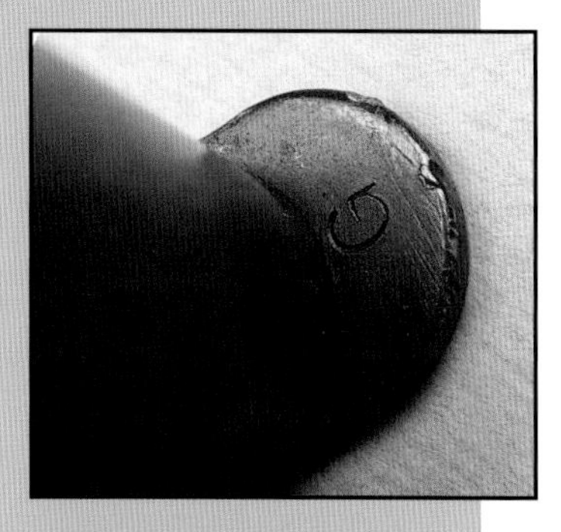

During the Second World War, a large number of subcontractors were called on to produce magazines. *Marc de Fromont*

- M. S. Little Co., Hartford, Connecticut (*C-R* or *R* marking)
- Ridson Mfg. Co., Naugatuck, Connecticut (*C-R* or *R* marking)
- Scovil Mfg. Co., Waterbury, Connecticut (*C-S* or *S* marking)

- **Hammer:** the Colt M1911A1 had been mounted with a wide cocking piece, though some hammers had a narrow cocking piece with a grid pattern or striations. They were made by Colt, Gray Mfg. Co. (Hartford, Connecticut), Wright Engineering Co., and Remington Rand.

- **Trigger:** the first models, which were machined, had a trigger bow of equal sections on the three sides. This was substituted with a trigger with a bow made of pressed steel, whose rear side has a small indentation to eliminate the risk of getting caught up in the trigger. This part was made by Yawman Metal Products, Rochester, New York.

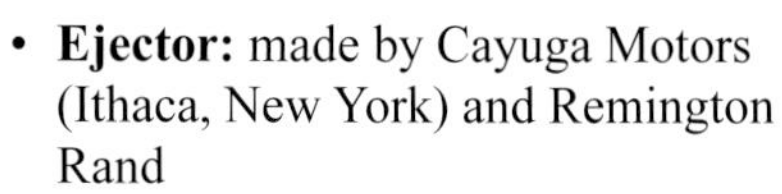

- **Ejector:** made by Cayuga Motors (Ithaca, New York) and Remington Rand

- **Barrel bushing:** production by Colt, Moore & Steele, Remington Rand Elmira, Remington Rand Ilion, and Wright Ingenierie Co.

- **Extractor:** part produced by Colt, Ithaca, and Remington

- **Stock screw bushing:** made by Colt and Hartford Screw Machine Co.

- **Trigger:** part made by Colt, Ithaca, and Remington Rand Elmira

- **Slides:** this part is struck with the name of the manufacturer, which had to correspond to the serial number.

- **Security handle:** part made by Colt, Ithaca, and Remington

- **Recoil spring housing:** made by Cayuga Motors (Ithaca, New York) and Remington Rand. The back of this part, smooth on the M1911, shows different finishes on the M1911A1.
  - Thin checkering grid pattern on Singer production and the older Remington production
  - Wide checkering grid pattern on the first Colt production as well as on Ithaca weapons and Union Switch and Signal
  - Vertical striation (seven for Colt and Ithaca, eight for Remington)

OPPOSITE PAGE: the Colt M1911A1 was the regulation weapon of the Korean Battalion, equipped entirely by the Americans. *Col. A. Nowak/ Marc de Fromont*

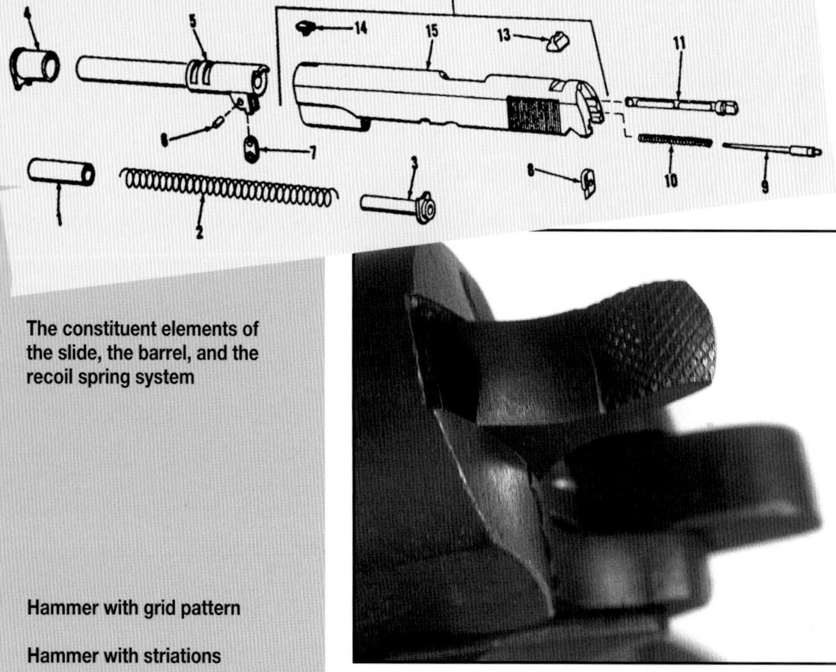

The constituent elements of the slide, the barrel, and the recoil spring system

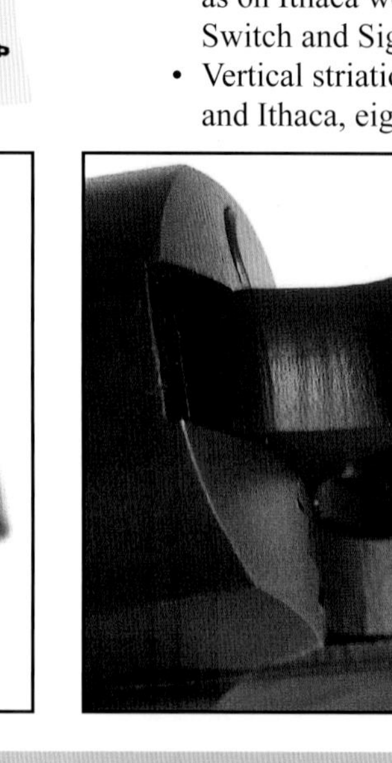

Hammer with grid pattern

Hammer with striations

UNITED STATES PROPERTY
№921547
M1911A1 U.S.ARMY
COREE
O U

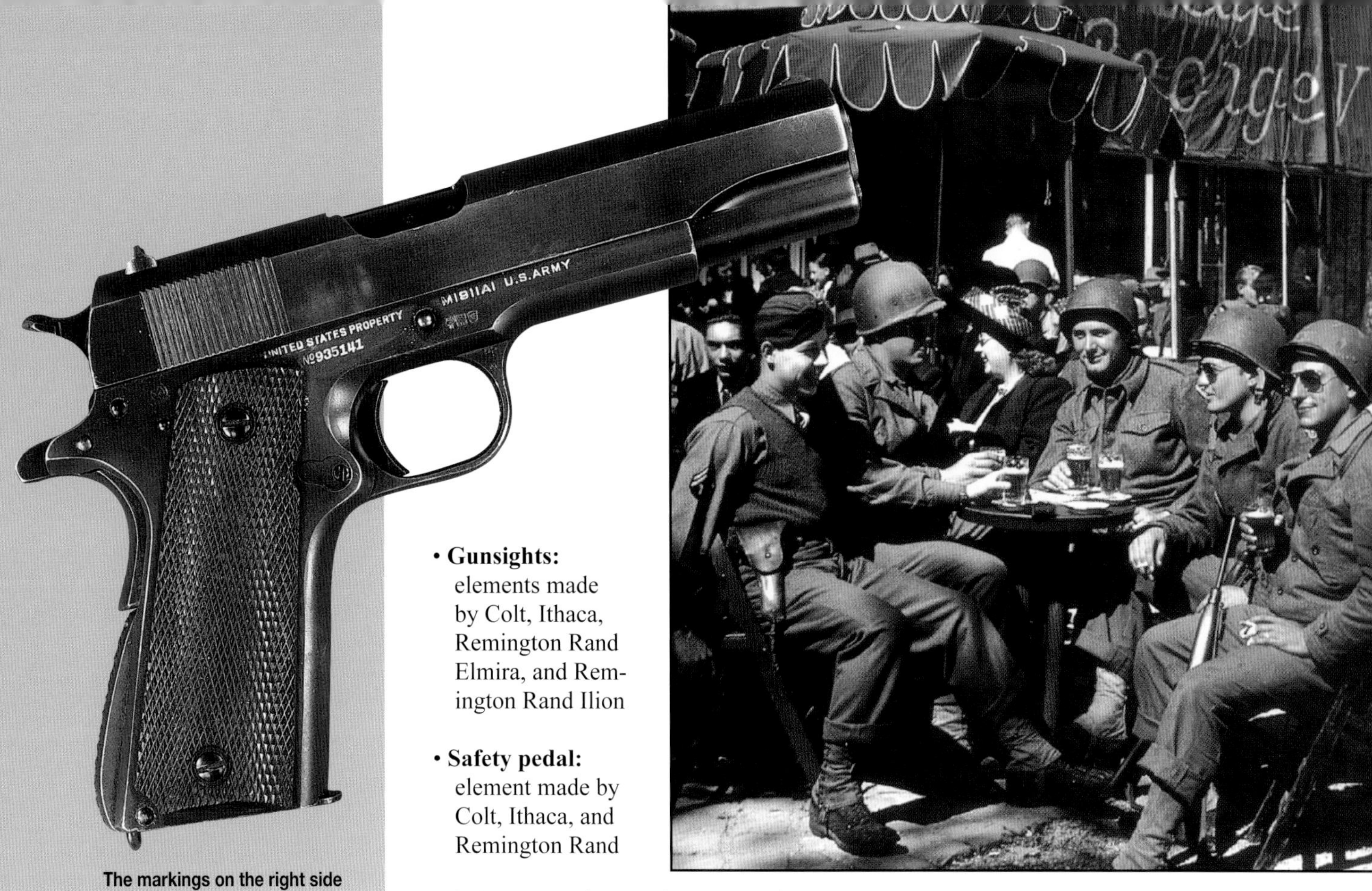

The markings on the right side of a Colt M1911A1

American soldiers at a street café in Paris, 1944. *US National Archives*

- **Gunsights:** elements made by Colt, Ithaca, Remington Rand Elmira, and Remington Rand Ilion

- **Safety pedal:** element made by Colt, Ithaca, and Remington Rand

- **Grip plates:** the exterior aspect of the grips was practically unchanged, but the internal side could be hollow, with or without reinforcement. They were produced by Colt or Keyes Fibre in New York.

- **Magazine release catch:** parts produced by Colt, Hartford Screw Machine Co., Remington Rand Elmira, and Remington Rand Ilion

- **Springs:** the majority were produced by Humason Mfg. Co., Forestville, Connecticut

- **Divider:** part made by Colt, Ithaca, and Remington

- **Recoil spring guide:** made by Colt, Moore & Steele, and Remington Rand

- **Magazine catch:** part made by Colt, General Pressed Metal Co. (Syracuse, New York), Remington Rand Elmira, and Remington Rand Ilion

Markings

Apart from moments of hesitation at the beginning of manufacture, the M1911A1 pistols bore the following markings on their constituent elements.

On the left of the slide, the dates of the principal patents, the business name, and the address of the manufacturer, which on the military Colt (with the logo of the colt between the dates and the business name) were

PATENTED APR. 20. 1917. SEPT 9 1902 –
COLT'S P.T. F.A. MFG. CORP
DEC. 19 1905. FEB. 14. 1911. AUG. 19.
1913 – HARTFORD. CT. U.S.A.

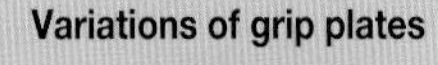
Variations of grip plates

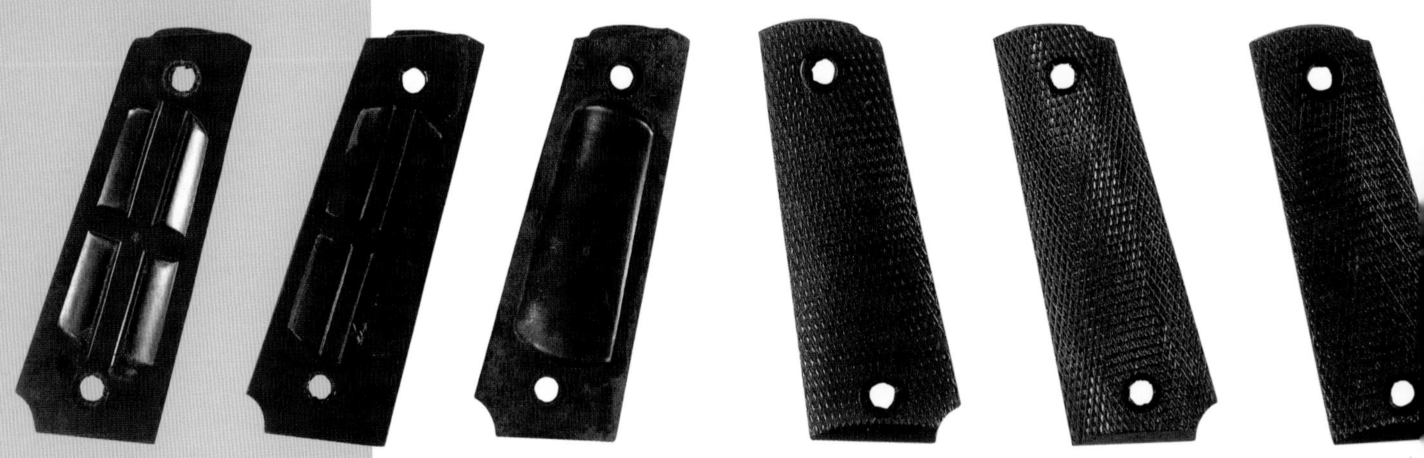

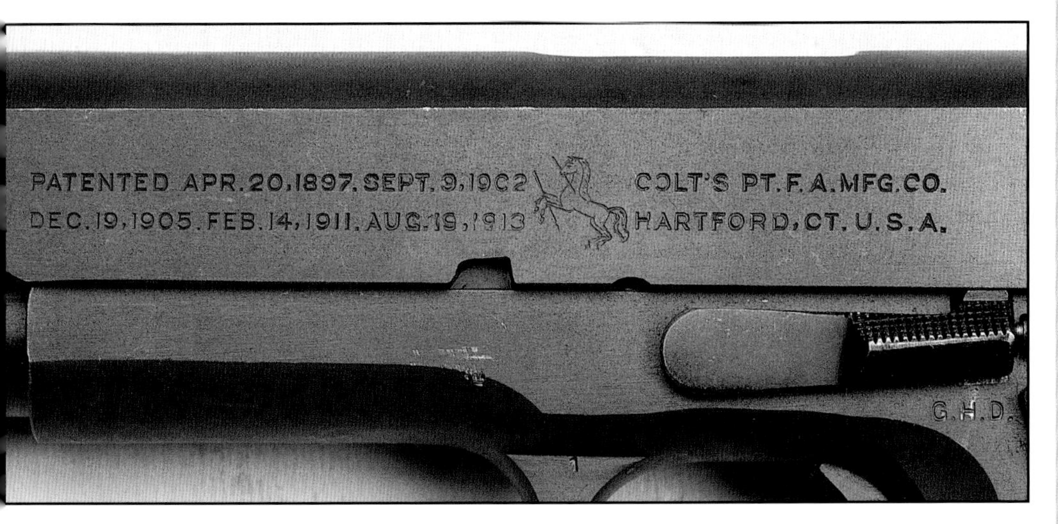

Markings on the slide on a Colt-produced M1911A1.
Marc de Fromont

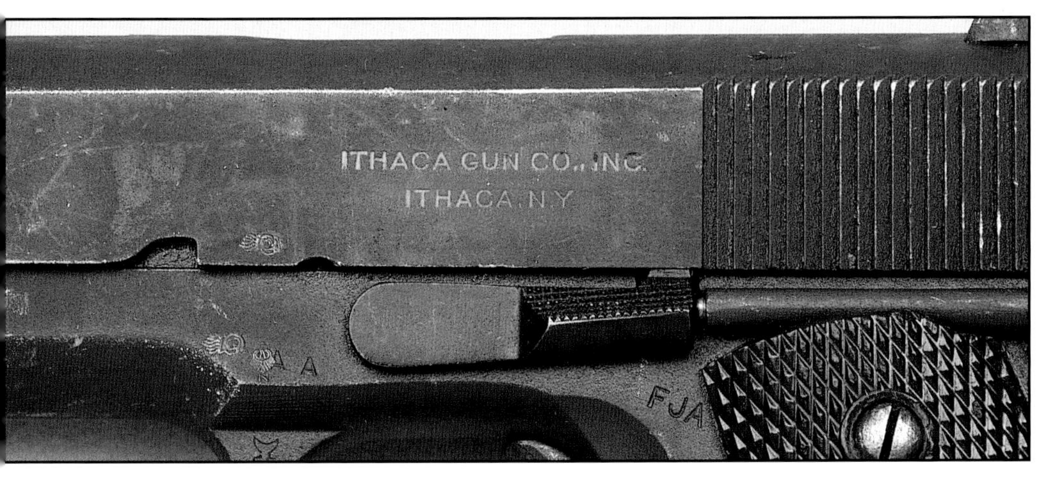

Markings on the slide on an M1911A1 produced by Ithaca.
Marc de Fromont

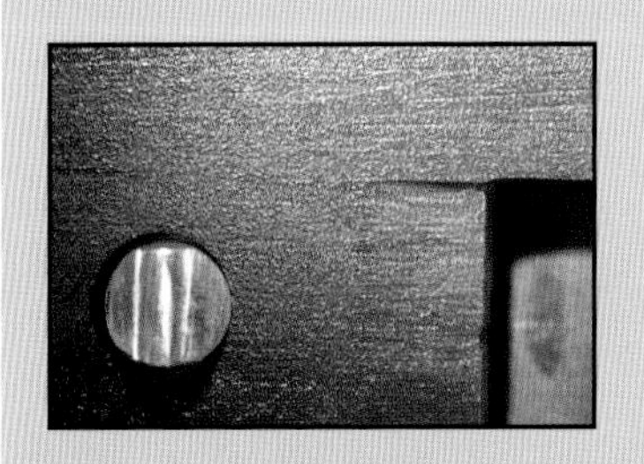

Absence of markings on the body of a weapon produced by Ithaca

On commercial Colts reconverted into military models, the markings (minus the colt) were

COLT'S PT. F.A. MFG. CO. HARTFORD, CT.
U.S.A.
PAT'D APR 20. 1897, SEPT 2. 1902, DEC 19
1905, FEB 14, 1911, AUG 19. 1913

For Singer:

S. MFG. CO.
ELIZABETH, N.J., U.S.A.

For Ithaca:

ITHACA GUN CO. INC.
ITHACA N.Y.

For Remington, a 60 mm long marking is found on the first makes:

REMINGTON RAND, INC.
SYRACUSE, NEW YORK

This was then replaced by a simplified marking measuring 35 mm; this transition took place between the numbers 935000 and 955000:

REMINGTON RAND, INC.
SYRACUSE, N.Y. U.S.A.

Following on, supposedly, from a change in equipment, an identical but smaller marking (25 mm approximately) was applied from a section of numbers between 980000 and 1015000. For Union Switch & Signal Company, the logo of the company (a large *U* containing two *S*'s), followed by

US&S CO.
SWISSVALE, PA. U.S.A.

On the right of the slide, on the commercial Colts reconverted into military models:

COLT Automatic
Caliber .45

The marking is followed by the Colt company logo.

Military pistols bore no marking on the right side of the slide.

On the right of the body:

UNITED STATES PROPERTY – M1911A1 U.S.
ARMY

Followed by the serial number and preceded by the symbol NO are the exception to this rule for the weapons produced by Remington from (approximately) no. 1000000, which bear the symbol NO.

OPPOSITE PAGE: the M1911A1 remained in service in the French army until at least the end of the war with Algeria. *Col. A. Nowak / Marc de Fromont*

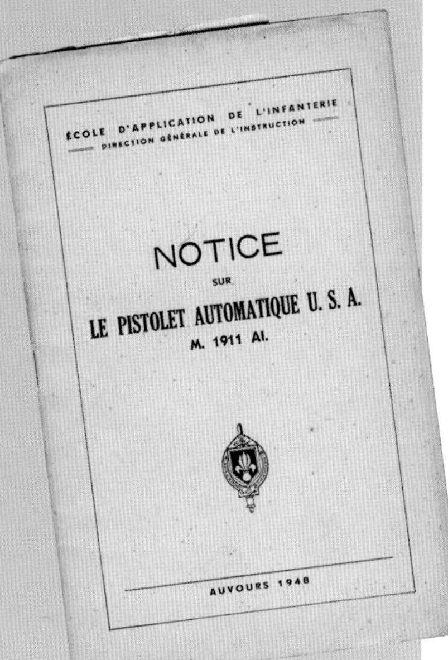
ÉCOLE D'APPLICATION DE L'INFANTERIE
DIRECTION GÉNÉRALE DE L'INSTRUCTION

NOTICE

SUR

LE PISTOLET AUTOMATIQUE U. S. A.

M. 1911 Al.

AUVOURS 1948

Colt 1911A1 handbook for the infantry school, 1948

Inspection Stamps and Other Markings

As for the M1911 pistols destined for the armed forces, the M1911A1 received the mark of the military inspector on the left side of the body, under the slide stop.

On weapons produced by Colt between 1924 and 1937, the stamp of Maj. Walter T. Gordon, in a circle, is present.

Between 1937 and 1942, Maj. Charles S. Reed was in charge of inspection. His stamp is simply his initials on a single line: CSR.

Due to the war production, other inspectors intervened, with the following stamps:

R.S.: for Col. Robert Sear (Colt)

W.B.: for technical officer Waldemar Broberg (Colt)

J.S.B.: for Col. John S. Bigley (Colt)

JCK: for Col. John K. Clement (Singer)

F.J.A.: for Col. Frank J. Atwood (Ithaca and Remington Rand)

R.C.D. in a circle: for Lt. Col. Robert C. Downie (Union Switch & Signal)

In addition on the left side of the body, there is a test stamp *P*, seen under the magazine catch.

Preinspection stamps were put on the body next to the divider housing. These stamps disappeared in October 1942 and were replaced by the logo of the ordnance (two crossed barrels joined in a circle and mounted on a grenade). This logo is stamped on the right side of the body, at the top and just behind the grip plate. It appears on all pistols produced by Colt (after no. 830000), Remington, and Ithaca, but not on those produced by Union Switch & Signal.

Markings on the slide of an M1911A1 produced by Remington

Markings on the slide of an M1911A1 produced by Union Switch & Signal. *Marc de Fromont*

ITHACA GUN CO., INC.
ITHACA N.Y.
ALGERIE
REPUBLIQUE FRANÇAISE

Inspection stamp of the Inspector Gen. Guy H. Drewry

Inspection stamp of Col. Frank J. Atwood

Stamp on the body (*left side*)

Stamp on the body (*right side*)

OPPOSITE PAGE: Still used in Vietnam, the M1911A1 chalked up many years of service. *Col. A. Nowak / Marc de Fromont*

Marking seen on some barrels (*left side*)

1943 manual for the French army

Before entering a tunnel during the Vietnam War. *US National Archives*

The magazines made by Colt were not marked; the other manufacturers or subcontractors had specific markings on the magazine base:

C-L or L for M. S. Little Co.

C-R or R for Ridson Mfg. Co.

C-S or S for Scovil Mfg. Co.

G for General Shaver Division

As a general rule, the barrels carry markings enabling the identification of their origin:

- Those made at Colt are marked COLT 45 AUTO on the left side or the top; the letter *P* is also found on the left side of the base or there is an encircled *C* on the right side, or both.
- The barrels made at the Springfield arsenal are marked with the letters *SP* on the left side of the base.

VIETNAM
50 CARTRIDGES
PISTOL BALL
CALIBER .45 M1911
AMMUNITION LOT E.C. 3-S-25292
UNITED STATES PROPERTY
№ 1059514
M 1911 A1 U.S. ARMY
REPUBLIC OF VIETNAM SERVICE

French information sheet of the 1950s

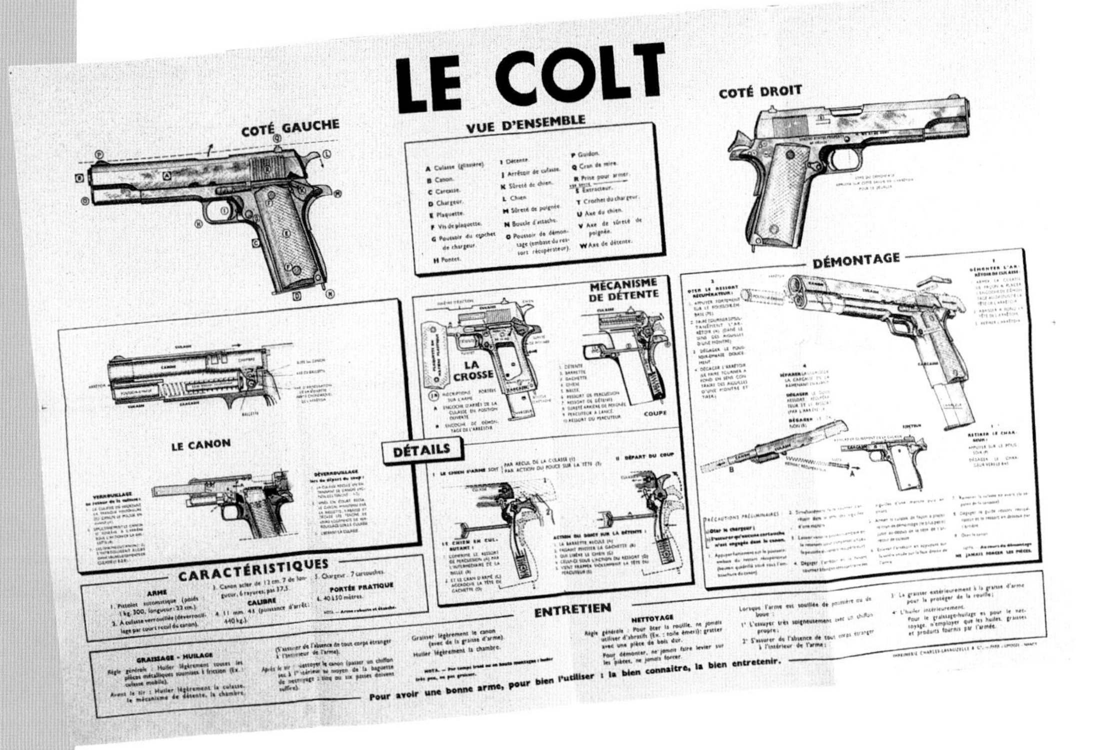

LE COLT

COTÉ GAUCHE

COTÉ DROIT

VUE D'ENSEMBLE

A Culasse (glissière).
B Canon.
C Carcasse.
D Chargeur.
E Plaquette.
F Vis de plaquette.
G Poussoir du crochet de chargeur.
H Pontet.
I Détente.
J Arrêtoir de culasse.
K Sûreté de chien.
L Chien.
M Sûreté de poignée.
N Boucle d'attache.
O Poussoir de démontage (embase du ressort récupérateur).
P Guidon.
Q Cran de mire.
R Prise pour armer.
S Extracteur.
T Crochet du chargeur.
U Axe du chien.
V Axe de sûreté de poignée.
W Axe de détente.

LE CANON

LA CROSSE

MÉCANISME DE DÉTENTE

DÉMONTAGE

DÉTAILS

CARACTÉRISTIQUES

ARME
1. Pistolet automatique (poids 1 kg. 300, longueur : 22 cm.)
2. A culasse verrouillée (déverrouillage par court recul du canon).
3. Canon acier de 12 cm. 7 de longueur, 6 rayures; pas 37,5.

CALIBRE
4. 11 mm. 43 (puissance d'arrêt : 440 kg.).
5. Chargeur : 7 cartouches.

PORTÉE PRATIQUE
6. 40 à 50 mètres.

ENTRETIEN

GRAISSAGE - HUILAGE
Règle générale : Huiler légèrement toutes les pièces métalliques soumises à friction (Ex. : culasse mobile).
Avant le tir : Huiler légèrement la culasse, le mécanisme de détente, la chambre.
(S'assurer de l'absence de tout corps étranger à l'intérieur de l'arme).
Graisser légèrement le canon (avec de la graisse d'arme).
Huiler légèrement la chambre.

NETTOYAGE
Règle générale : Pour ôter la rouille, ne jamais utiliser d'abrasifs (Ex. : toile émeri); gratter avec une pièce de bois dur.
Pour démonter, ne jamais faire levier sur les pièces, ne jamais forcer.

Lorsque l'arme est souillée de poussière ou de boue :
1° L'essuyer très soigneusement avec un chiffon propre;
2° S'assurer de l'absence de tout corps étranger à l'intérieur de l'arme;

Pour avoir une bonne arme, pour bien l'utiliser : la bien connaitre, la bien entretenir.

Reception stamp for barrels

- The barrels made by Remington are struck with a *P* on the upper part of the tube.
- A *P* marking is found on the left side of the base on barrels produced by Singer, but barrels with no markings also exist.
- Barrels made by High Standard bear a *P* on the left side of the base, and the letters *HS* on the right side of the barrel.
- Those produced by Flannery Bolt Co. bear a *P* on the left side of the base, and the letter *F* on the right side of the barrel.

A medic of the 25th US Infantry Division takes care of a wounded soldier during the Vietnam War. *US Army*

The Resumption of Production

Colt resumed production of 1911A1 models in October 1946 for the civilian sector with the no. C220001.

The conception of the weapon remained unchanged until 1970.

CHAPTER 5

IMPLEMENTATION

The M1911A1 has a breech latch.

Before using any type of semiautomatic pistol, it is necessary to carry out a certain number of maneuvers.

In the case of the Colt M1911 and M1911A1 pistols, it is necessary to push the magazine support catch, situated on the left side at the base of the trigger guard.

Under the action of its spring, the magazine is released automatically from its housing, and a follow-through movement by the hand is necessary to extract it.

The seven-cartridge magazine is filled easily without the use of a tool.

The magazine is replaced by putting it into its housing until it locks into position. This operation is called inserting.

Operation

The Colt .45 pistol is a powerful weapon. It is capable of propelling a 15-gram bullet at a speed greater than 560 mph (900 km/h).

This results in heavy constraints and, notably, a raised pressure inside the bullet casing, making the cartridge have an impact on the mechanism. The Colt pistol belongs therefore to the category of short-breech weapons, but in order for the automatic repetition mechanism to function, it must call on a principle known as short barrel recoil.

Since the weapon had been loaded beforehand, the action of pulling the slide completely toward the rear would cause the mechanism to load, then the introduction of a cartridge in the chamber when the slide returns forward.

The weapon is now loaded.

The user must hold the weapon firmly, pushing down the safety lock at the same time as squeezing the trigger. The latter will lead to certain parts moving: first the trigger group, then the trigger, which as it retracts frees the hammer.

When the hammer is lowered it gives a forward impetus to the firing pin, which strikes the primer of the cartridge, and firing takes place.

The powder of the cartridge is consumed at a very great speed, releasing gases. As the gases expand, equal pressure is exerted in all directions. These gases force the bullet down the barrel and force the slide to the rear. At the same time, pressure is exerted on the base of the cartridge; as a result of this pressure the barrel and slide unit move rearward. The barrel and slide move together by the housing of the studs, situated on the upper part of the barrel in the corresponding openings on the slide.

The weapon has a single-action trigger that necessitates advance arming of the hammer before firing. *US National Archives*

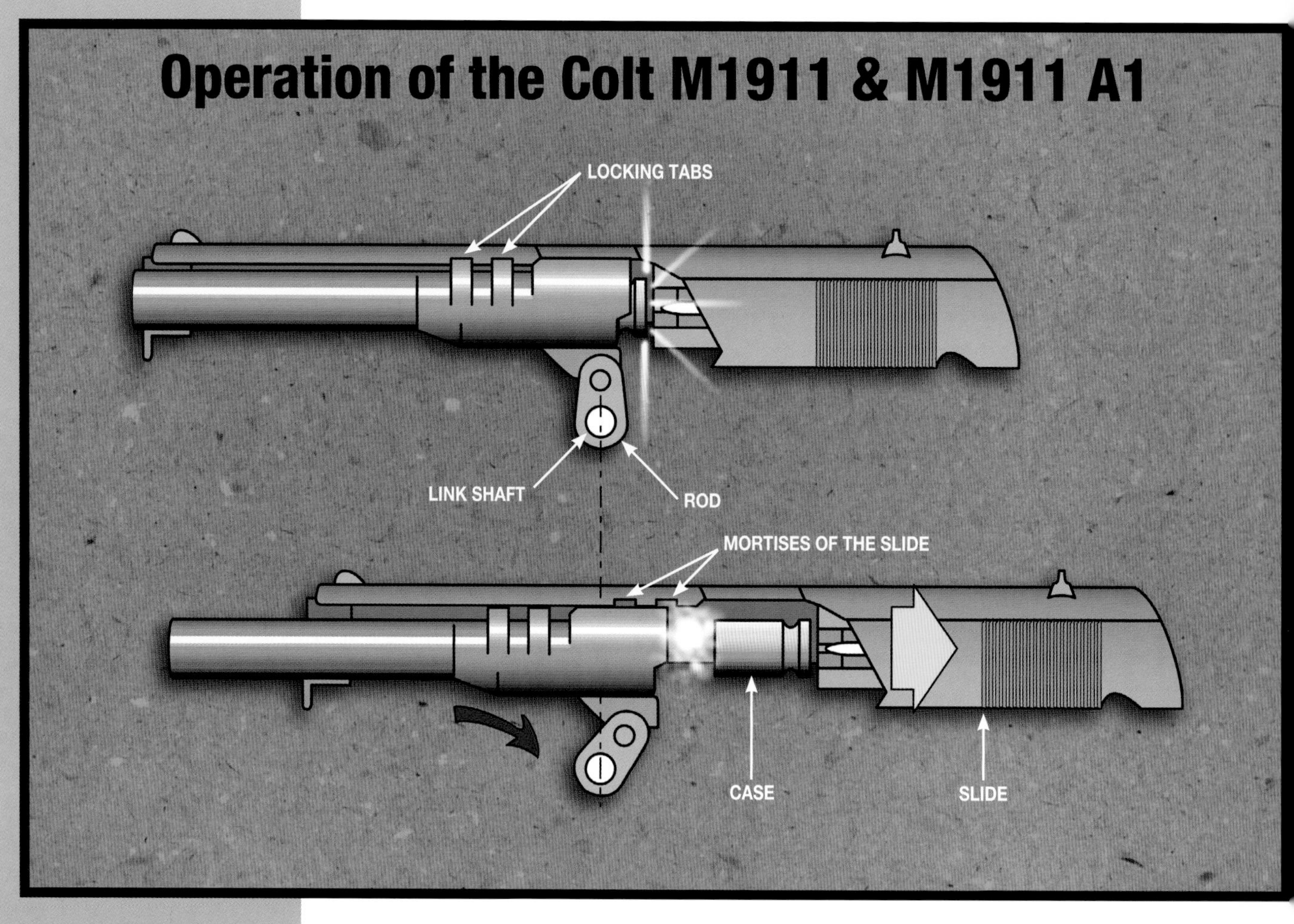

The weapon is operated by short recoil of the barrel with a fixed breech when the shot is fired.

But the lower part of the barrel is linked by means of an articulated rod on the axis of the latch, which limits its movement. The rear part of the barrel is lowered and the tube ends its movement here. The slide, divided from the barrel, continues its maneuver. With the extraction and ejection of the bullet casing, the slide finishes its rear movement by creating the rearming of the hammer, while the recoil spring is compressed.

The spring makes the slide return forward, and during this movement it meets a new round at the top of the magazine. This round is pushed into the chamber. The slide unites with the barrel (closing), and the unit returns to a forward position (locking). The forward movement is finished.

This cycle is repeated every time the user puts pressure on the trigger, until the magazine is empty.

When the magazine is empty, the magazine follower makes the breech latch move upward and immobilizes the slide in a rear position.

The fact of replacing an empty magazine by a full one makes the latch retract by allowing the breech to close and the pistol to reload.

If for some reason the slide did not come back to the forward position completely, a part of the mechanism, called the divider, intervened in the kinematic chain of the elements, which causes the bullet to be fired and stops any premature firing or possible rapid fire.

When the pistol is loaded it is also possible to engage the optional safety, which is on the left rear of the body. This movement has the effect of immobilizing the slide and the hammer. However, for safety reasons, this practice is not advised unless for operational use of the weapon under extreme conditions.

Throughout its lifetime the M1911A1 has been legendary, and few soldiers remained indifferent when faced with such a weapon.

CHAPTER 6

DISASSEMBLY OF THE COLT 1911A1

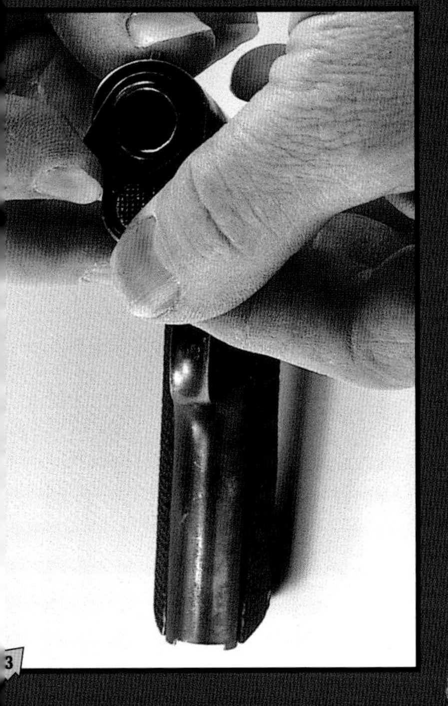

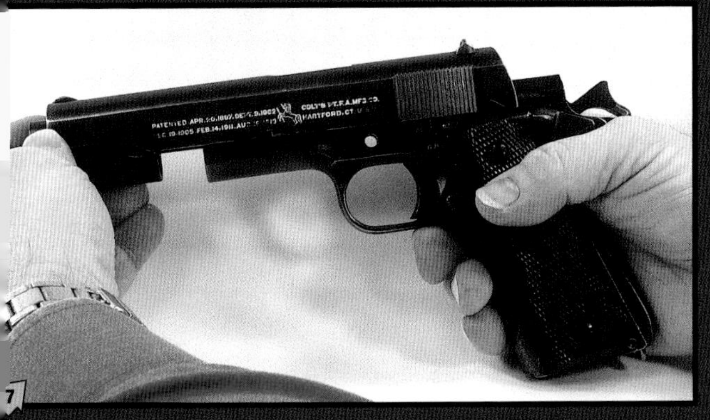

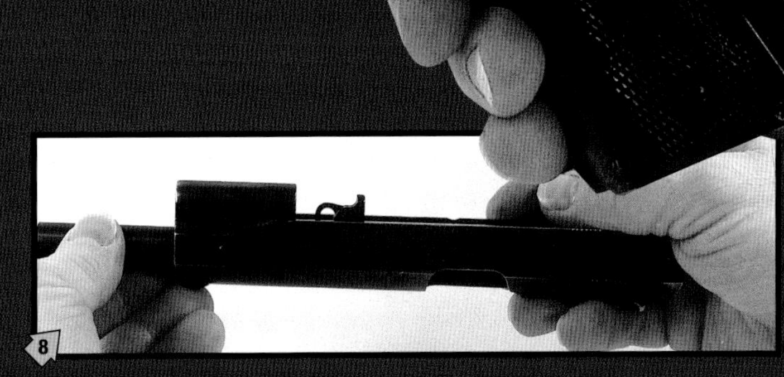

1. Remove magazine by pressing the bolt situated on the left, just behind the emergence of the trigger.

2. As a safety measure, maneuver the slide two times to ensure that no cartridge remains in the chamber.

3. Depress the recoil spring plug and turn the barrel bushing a quarter turn to the right. The plug and recoil spring can then be extracted from their housing.

4. Turn the bushing slightly less than a half turn to the right; this permits the disassembly.

5. Bring the slide to the rear until the top of the slide lines up with the smallest notch on the bottom left of the slide.

6. Press the slide stop into frame, and the slide stop will come free.

7. Move the slide forward to separate the slide from the frame.

8. Push the barrel out of the front of the gun, having taken care to move the connecting rod downward.

9. The disassembly of the other parts can now continue, in theory without tools. These operations are outside the scope of this study.

Reassembly

Reassembly is carried out in reverse order to disassembly. After having put the barrel, recoil spring, and its buffer pitot back into place, this unit is mounted on the frame and the slide stop is positioned.

It is then necessary to engage the manual safety before positioning the barrel bushing (its extension is on the left) and to spin it approximately 175 degrees to the right.

It remains to position the recoil spring plug, then move back the barrel bushing.

CHAPTER 7

SERIAL NUMBERS

MODEL	YEAR	SERIAL NUMBER	MANUFACTURER	NOTE	PRODUCTION
M1911	1912	1–500	Colt		500
		501–1000	Colt	USN	500
		1001–1500	Colt		500
		1501–2000	Colt	USN	500
		2001–2500	Colt		500
		2501–3500	Colt	USN	1,000
		3501–3800	Colt	USMC	300
		3801–4500	Colt		700
		4501–5500	Colt	USN	1,000
		5501–6500	Colt		1,000
		6501–7500	Colt	USN	1,000
		7501–8500	Colt		1,000
		8501–9500	Colt	USN	1,000
		9501–10500	Colt		1,000
		10501–11500	Colt	USN	1,000
		11501–12500	Colt		1,000
		12501–13500	Colt	USN	1,000
		13501–17250	Colt		3,750
	1913	17251–36400	Colt		19,150
		36401–37650	Colt	USMC	1,250
		37651–38000	Colt		350
		38001–44000	Colt	USN	2,000
		44001–60400	Colt		16,400
	1914	60401–72570	Colt		12,170
		72571–83855	Springfield Armory		11,285
		83856–83900	Colt		45
		83901–84400	Colt	USMC	500
		84401–96000	Colt		11,600
		96001–97537	Colt		1,537
		97538–102596	Colt		5,059
		102597–107596	Springfield Armory		5,000
	1915	107597–109500	Colt		1,904
		109501–110000	Colt	USN	500
		110001–113496	Colt		3,496
		113497–120566	Springfield Armory		7,070
		120567–125566	Colt		5,000
		125567–133186	Springfield Armory		7,620
	1916	133187–137400	Colt		4,214
	1917	137401–151186	Colt		13,786
		151187–151986	Colt	USMC	800
		151987–185800	Colt		33,814
		185801–186200	Colt	USMC	400
		186201–209586	Colt		23,386
		209587–210386	Colt	USMC	800
		210387–215386	Colt		5,000
		215387–216186	Colt	USMC	800
		216187–216586	Colt		400
		216587–216986	Colt	USMC	400

Colt and Springfield produced more than 580,600 pistols during the First World War. *Jean-Claude Fombaron*

MODEL	YEAR	SERIAL NUMBER	MANUFACTURER	NOTE	PRODUCTION
	1918	216987–217386	Colt	USMC	400
		217387–232000	Colt		14,614
		232001–233600	Colt	USN	1,600
		233601–580600	Colt		347,000
		1– 3152	Remington UMC		13,152
	1919	13153–21676	Remington UMC		8,524
		580601–629500	Colt		48,900
		629501–700000	Colt	Order suspended	0
M1911*	1924	700001–711605	Colt		10,000
M1911A1	1937	710001–711605	Colt	Army and USN	1,605
		711606–712349	Colt	USN	744
	1938	712350–713645	Colt		1,296
	1939	713646–717281	Colt	USN	3,636
	1940	717282–721977	Colt		4,696
	1941	721978–756733	Colt		34,756
	1942	756734–793657	Colt		36,924
		793658–797639	Colt	USN	3,982
		797640–800000	Colt		2,361
		800001–856100	Colt		55,100
	1943	S 800001–S 800500	Singer		500
		H 800501–H 801000	Harrington & Richardson		0
		856101–958100	Colt		102,000
		856101–856404**	Renumbered		304****
		856405–916404**	Ithaca		60,000
		916405–1041404**	Remington Rand		125,000
		1041405–1096404**	Union Switch & Signal		55,000
		1088726– 092896**	Colt		4,171
		1096405–1208673	Colt		112,269
		1208674–1279673	Ithaca		71,000
		1279674–1279698	Renumbered		25****
		1279699–1441430	Remington Rand		161,732
		1441431–1471430	Ithaca		30,000
		1471431– 609528	Remington Rand		138,098
	1944	1609529–1743846	Colt		134,318
		1743847– 816641	Remington Rand		72,795
		1816642–1890503	Ithaca		73,862
		1890504–2075103	Remington Rand		184,600
	1945	2075104–2134403	Ithaca		59,300
		2134404–2244803	Remington Rand		110,400
		2244804–2380013	Colt	2,368,781***	123,978
		2380014–2619013	Remington Rand	2,465,139***	85,126
		2619014–2693613	Ithaca	2,660,318***	41,305
TOTAL estimated production of Colt M1911 and M1911A1:					2,551,730

USN = US Navy
USMC = US Marine Corps

* M1911 improved (model of transition between the M1911 and M1911A1)
** Numbers included in the sections having a dual number
*** The highest serial number made in the series
**** Not counted in the total production

The M1911A1 was also present in the Resistance, as seen here during the liberation of San Tropez.

1. Serial number of a pistol made by Ithaca in 1945
2. Serial number of a pistol made by Union Switch & Signal in 1943
3. Serial number of a pistol made by Remington in 1943

Ever since specialist authors, historians, and collectors have been studying Colt M1911 and 1911A1 pistols, there has never been any agreement on the exact quantity of weapons produced or on the serial numbers attributed to the units that used the weapons or their manufacturers!

Why Such a Variety of Opinions?

Reseachers have used official archival documents accessible to the general public, or at least those that the manufacturers wanted to make public.

Even if all involved grasped the difficulty of the problem in a general way, there remains a distortion between the serial numbers assigned in theory and those that are encountered in reality.

This is explained by the fact that various contracts were canceled before they were able to be fully carried out, and the unused numbers were not taken up administratively at a later date.

In order to simplify things and following on from the errors of the ordnance, some numbers were assigned twice! It would be relatively easy to clarify things if there were only two manufacturers concerned, but there were many more.

The correlation that exists between the groups of numbers assigned to a manufacturer and the year of production is also subject to discussior depending on whether the reference is

- the year the order was placed (generally corresponding to the tax year for which the batch of weapons was budgeted), or
- the year when delivery was carried out,

in addition to the various "slippages" that took plac

Nonetheless, the most-credible figures fron the most-recent sources on the matter will b given here.

These numbers concern only military models for information on commercial models, pleas refer to the relevant chapters.

Double Numbering

After Pearl Harbor, the United States was throw into an economic crisis that the militar administration was not always able to keep u with. As a result, the ordnance, tasked wit assigning serial numbers to the makers of th M1911A1 pistol, mistakenly allocated the sam group of numbers to different manufacturers: i is estimated that 105,867 examples of the M1911A pistol bore duplicated numbers:

- 60,000 Colt-made pistols in the section of numbers reserved for Ithaca
- 41,696 others in the section reserved for Remington Rand
- 4,171 in the section reserved for Union Switch & Signal

The same mistake occurred with the M1 Garand made by Springfield Armory and Winchester.

Weapons of Series X

Furthermore, the arsenals were restoring a certai number of pistols. The exact number of M1911 repaired in this way is not known because th operation was extensive, beginning in 1924, an continuing until 1957 with the M1911A1. W know only that an approximate total of 5,00 pistols were made usable again.

In theory, weapons repaired in the arsenal bear special markings that are unique to eac establishment:

- *AA* for Augusta Arsenal
- *RIA* for Rock Island Arsenal
- *SA* for Springfield Armory

CHAPTER 8

VARIATIONS OF THE COLT

and Weapons Modified by the Arsenals

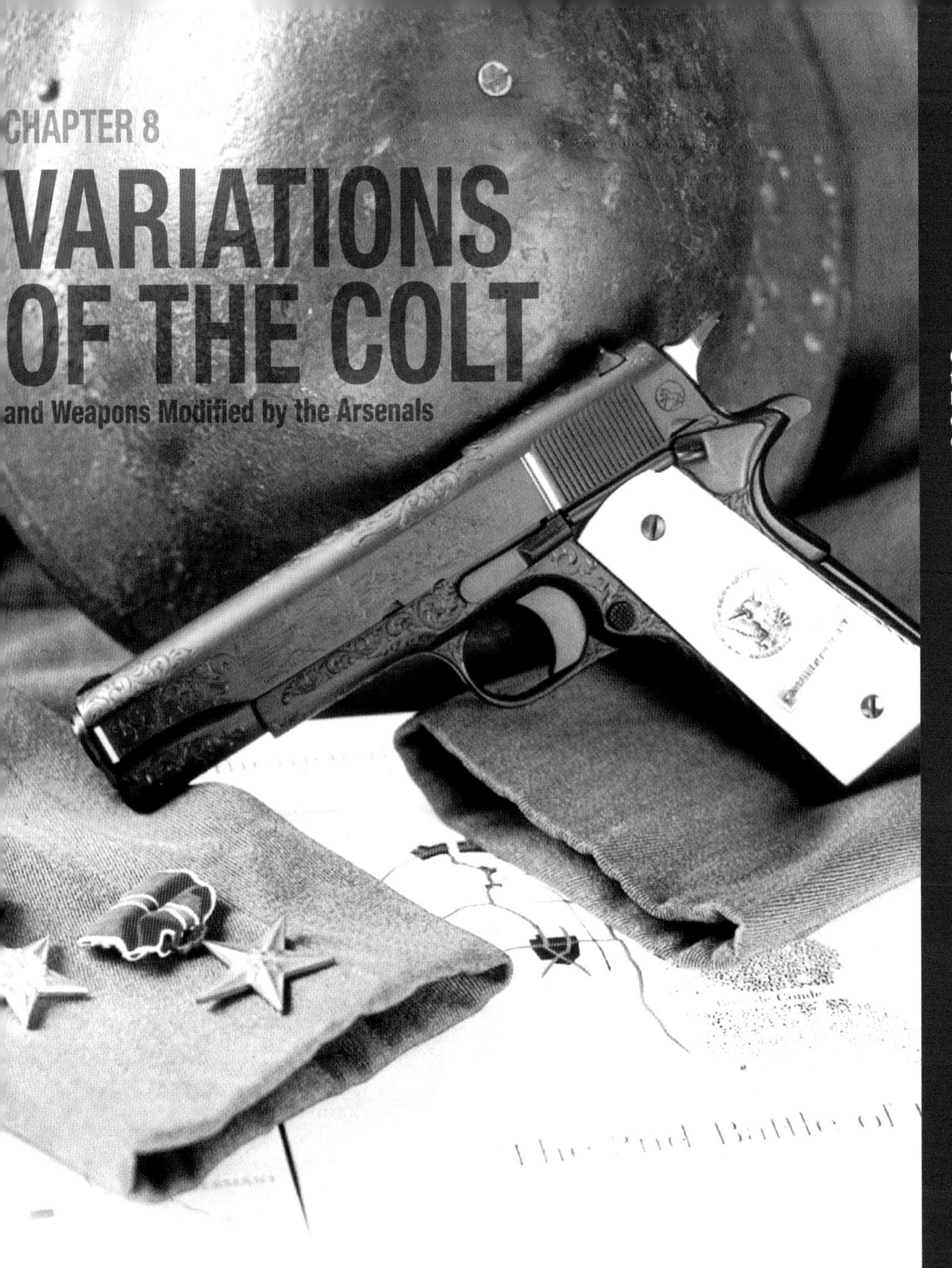

Commemorative Colt on the theme of great battles of the First World War. This one is dedicated to the second battle of the Marne.

Standard Commercial Weapons

Starting in 1912, the Colt company sold regulation-model (M1911, then M1911A1) pistols in the civilian sector alongside its military production. These weapons bore a serial number preceded by the letter *C*. This production was suspended in 1942 with no. C215018.

It was taken up again in 1946 with no. C221001. From 1950 onward, the method of numbering was modified slightly; the letter *C* came *after* the serial number (around no. 240228C). The production of the commercial Government Model ended in 1969 with no. 33619C.

Weapons produced after the war were initially assembled with walnut grip plates with grid pattern, then substituted with plastic grip plates with the manufacturer's logo, then again with walnut grip plates with a medallion.

Colt offered their pistols with a bronzed or nickel-plated finish.

Engraved Weapons

Before the war, Colt offered three qualities of engraving: A, B, or C, with the latter being the highest level.

Since the end of the last world war, the Colt art workshop has offered to carry out any request that is submitted to them; Colt M1911 or M1911A1 have been created with special finishes, bronzed, or given a patina of silver, plated with precious metal or inlaid with gold to celebrate a special event or honor a celebrity.

Colt .38 Super Auto

Starting in 1928 and returning to its first passions (the Colt 1900 and 1902), the Colt company proposed a government model able to fire the .38 Super Auto round, which was more powerful than the .38 Auto.

The production of this model, of which 34,450 were made, was suspended in 1940. It recommenced in 1946 and continued until 1968, with around

Colt Ace Model .22, with a nonfixed breech. *Col. A. Nowak*

165,000 weapons made.

At the end of the war the American army bought 400 .38-caliber pistols from Colt. They were marked with the stamp G.H.D.

Colt Ace Model .22

The idea of a .22-caliber weapon taking on the appearance of a regulation pistol goes back to 1913. However, all attempts carried out by the arsenals, Colt, or other makers resulted in failure until 1931, when the manufacturer Hartfort proposed its Ace .22 model. This was a (semi) automatic pistol with the same dimensions and an identical weight to the Government Model, but firing .22 Long Rifle ammunition.

The weapon functioned with a nonfixed breech; its slide was lightened and fitted with a buffer system. There were no interchangeable parts between the Ace and the regulation weapon.

Just over ten thousand examples of this model were produced until 1941, and 752 of those were delivered to the armed forces.

Colt Ace Model .22 Service Model

Users of the Ace complained that there was no sensation of recoil when firing. In order to satisfy them, Colt made a weapon with a floating chamber that amplified the recoil energy. This was invented by David Marshal Williams, who made a similar device for reduced fire for the Browning machine guns. Williams was also the inventor of the US M1 carbine.

On sale from 1936 onward, these pistols have a serial number preceded by the letters *SM*. Out of a production of 12,460 guns up to 1945, 11,961 were delivered to military organizations and therefore bear the initials of the military inspection stamps: J.S.B. on the left side of the frame and UNITED STATES PROPERTY on the right side of the frame, above the serial number.

The manufacture of the Service Model was taken up again in 1978, but since the 1980s it no longer features in the catalog.

Kit .22/.45

This conversion, marketed in 1938, comprises

- a slide, with a trapezoidal foresight and a micrometric backsight adjustable both in height and direction,
- a recoil spring,
- a 5.5 mm caliber barrel with a floating chamber,
- a barrel bushing,
- an ejector, and
- a magazine.

COLT CAL. 22 L.R.

Mounted on a .45-caliber weapon, this conversion permitted the use of .22 Long Rifle cartridges.

It could also be fitted on a weapon chambered in .38 Super Auto or 9 mm, but in these cases the inner side of the ejector had to be slightly beveled so as not to hinder the functioning of the pistol in the original caliber.

The manufacture of this kit was suspended in 1946, after the production of 2,670 units (160 of which were supplied to the military). It was taken up again in 1978.

The parts of a Service Model .22 Ace, with a floating chamber

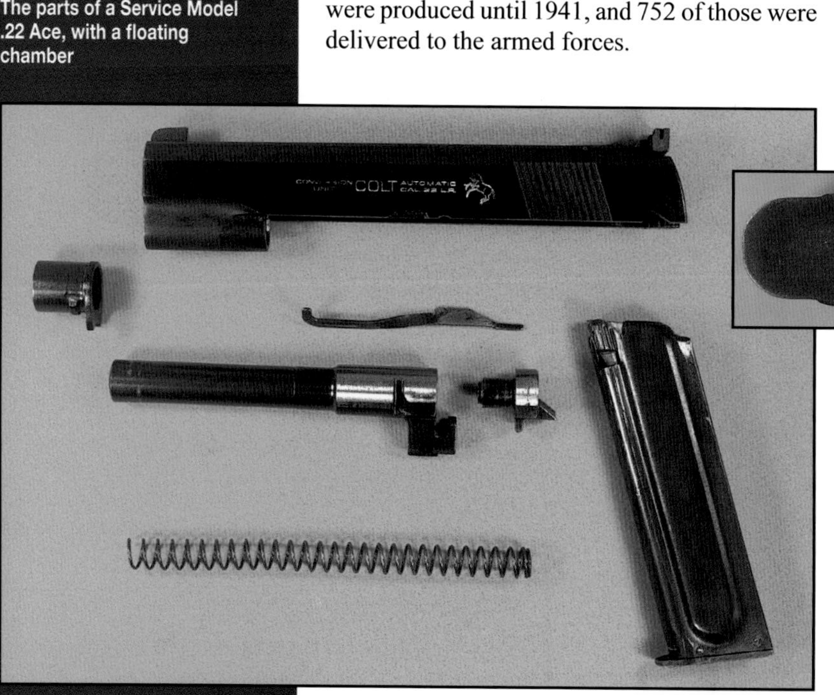

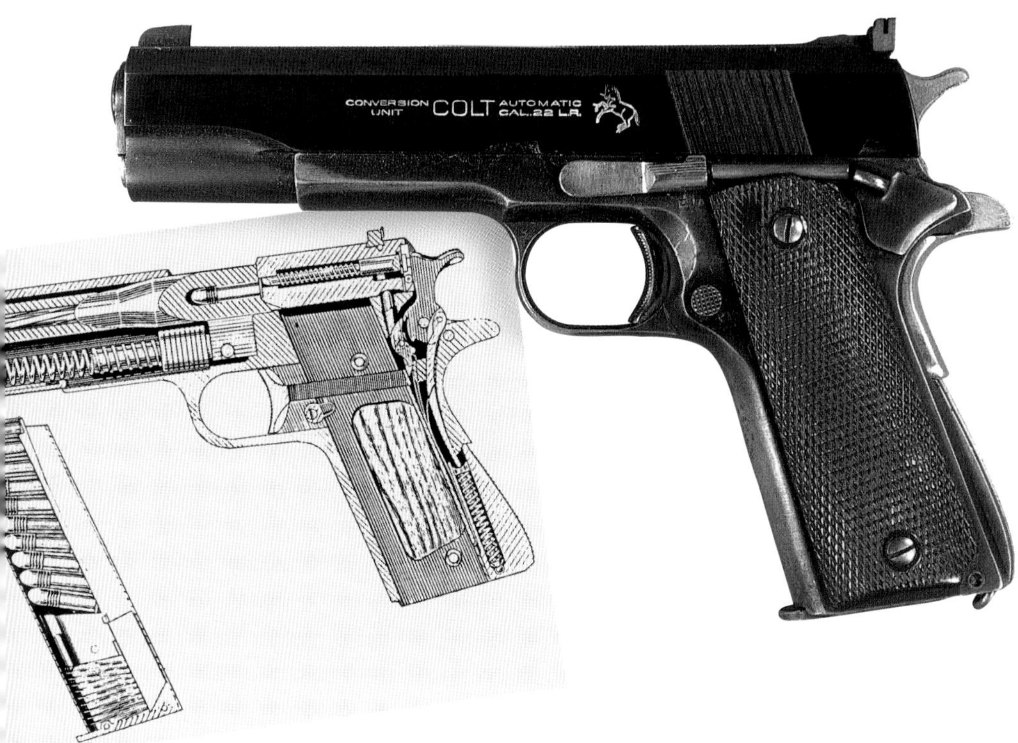

Service Model Ace .22 conversion mounted on an M1911A1 frame

Cross section of the Colt Ace .22 mechanism with nonfixed breech

Kit .45/.22

In 1932, the Colt company offered a version of its .45-caliber pistol in a Sports version. This model was marketed under the name National Match. It had a stone-polished mechanism, a barrel selected for its accuracy, an optional adjustable Stevens backsight, and a luxurious finish. Seven of these pistols were acquired by the US Coast Guard in 1932, and a further fifteen went to the US Marines in 1940.

Colt Super Match

The manufacturer also proposed a version equivalent to the .38 Super Auto, designated the Colt Super Match.

Colt Gold Cup National Match .45

In 1957, Colt recommenced the production of precision weapons with a lightweight aluminum trigger with an overtravel stop adjustment, a micrometric backsight, and a competition foresight. A new slide was fitted along with a larger ejection port, a new extractor, and wide oblique ridges for gripping; the upper part is flat and treated with antireflective coating.

The weapon received other improvements, notably a new barrel nut, which cut down on the slackness of the rod; in addition, the front side of the grip was striated and the hammer was redesigned.

Colt Gold Cup National Match .38

The .38 Super Auto was identical in shape to the Gold Cup .45. Between 1960 and 1971, 7,000 examples of this weapon were made.

Colt Gold Cup National Match Mid-Range

This more sports-oriented variation is chambered for the .38 Special Wadcutter cartridge. Apart from the technical innovations of the Gold Cup .45, it had a fixed barrel, the breech was not locked at the moment of fire, and a series of ribs in the chamber delayed the extraction and slowed down the recoil of the slide.

This model, also known under the name Mark III, was produced between 1961 to 1974.

Kits .38/.45

Colt marketed kits in Special Wadcutter .38, including the slide, the barrel, and the Mid-Range magazine, which could be mounted either on .45 ACP caliber or .38 Super Auto weapons.

These conversions were produced only during these date ranges:

- from 1964 to 1970; 434 examples for .38-caliber weapons
- from 1964 to 1969; 1,164 were made for .45 caliber and were tested by teams from the Advanced Marksmanship Unit; this is the origin of the name .38 AMU, which is how they are referred to on occasion.

In order to ensure a better locking of the barrel when the shot was fired, the Springfield Armory modified the locking system. The connecting rod is used only for unlocking, and the barrel puts weight on the lower part of the pin of the slide stop.

The most-recent weapons had a barrel where the dimensions of the chamber had been modified, along with a slide in carburized steel made by Drake Mfg. Co.

These weapons had a specific numbering system preceded by the letters *NM*, whereas the initials of the arsenal (*SA*) were struck on the right side of the frame, above the front brace of the trigger guard. The serial number of the weapon also appears on the barrel and its barrel bushing.

It is estimated that around 10,000 pistols were modified in this way by the army up until 1967.

.38 Special Wadcutter Conversion

Army National Match

In 1954, the US Army decided to equip its firing teams with an improved-performance regulation-model weapon. After fierce selection, 800 pistols were chosen to benefit from specific modifications. They were equipped with a new barrel and a new barrel bushing; the adjustment of the slide on the frame reduced the give during use, but the original sights were conserved. These pistols were provisioned at the Camp Perry firing range in 1955.

The year 1957 saw the addition of a raised sighting notch and an adjustable sight, while a neoprene lining was mounted on the front side of the grip. The following year, this model was improved again by the addition of a micrometric backsight. A trigger in plastic material came along in 1959, but this proved to be sensitive to greasy or fatty substances and was replaced in 1961 by a trigger in aluminum.

OPPOSITE: The Colt Commander is a commercial weapon that takes up the architecture of the .45 Service Model but is less bulky and uses 9 mm Parabellum rounds. ***Col. A. Nowak / Marc de Fromont***

Air Force Premium-Grade Match

Just like the army, between 1958 and 1969 the US Air Force modified a certain number of regulation pistols into high-level firing weapons designated AFPG. Later, the air force also used Colt National Match and Gold Cup, modified to their own criteria.

Colt Commander

In 1949, Colt proposed a lighter version of the Government Model, still in .45. The weapon is characterized by a 108 mm barrel and a shortened slide; the frame is in a light alloy coated in blue-toned black. The cocking piece is with grid pattern.

Colt Combat Commander

The Combat Commander is in every aspect identical to the Commander, except for the fact that it has a steel frame. The weapon is available in .45 ACP, .38 Super Auto, and 9 mm Parabellum.

Colt Commander. ***Marc de Fromont***

M15 General Officers' Pistol

From around 1973 onward, the Rock Island Arsenal modified a certain number of regulation pistols for its general officers; these modifications were inspired by the Colt Commander. The barrel and the slide are shorter, the front side of the grip has a grid pattern, and the weapon has a bronzed finish.

The grip plates are in walnut, with an encrusted grip on the left and the insignia of the ordnance on the right.

The markings that feature on this weapon, on the left of the slide, are

General Officer Model
RIA

The serial numbers are preceded by the letter GO on the right of the frame.

COLT
.45 AUTOMATIC
.38 AUTOMATIC
LUGER
of Magazine, 7 Shots for .45
9 Shots f
shot.
atically
COLT'S MFG. CO. HARTFORD CT. U.S.A.
4891-LW.

Colt commemorative pistol, 1911–1981, 3,000 of which were produced

Commemorative Weapons

In 1967 and 1968, Colt made a limited series of M1911 pistols commemorating the First World War. These weapons received a luxury treatment: polished bronzed finish, grips in walnut precious wood or ivory with insignia, and commemorative plate. On the right of the slide is a band marked

1917 – WORLD WAR ONE
COMMEMORATIVE – 1967

The other side is engraved with a battle scene recalling the principal American combats during this period. There are four variations, and 7,400 were made of each one (plus several others with a high luxury finish):

- Meuse-Agronne (serial number from 1 to 7400 followed by the letters *MA*)
- Bois-Belleau (serial number followed by the letters *BW*)
- Second Battle of the Marne (serial number followed by *M2*)
- Chateau Thierry (serial number followed by the letters *CT*)

On the same principle, Second World War commemorative weapons were also made in 1970. The model retained was with a nickel finish and smooth grip plates in precious wood with Colt insignia. There are two versions. Europe and African campaigns, with the following on the left of the slide:

WORLD WAR II COMMEMORATIVE
EUROPEAN THEATER OF OPERATION

And on the right side, the main interventions:

NORTH ATLANTIC – TUNISIA – SICILY
PLOESTI – SALERNO – ANZIO – NORMANDY
– BASTOGNE – REMAGEN – BERLIN

The serial number is followed by *ETO*.

Asia and Pacific campaigns, with the inverse markings in relation to the European version. The following is on the right side of the slide:

WORLD WAR II COMMEMORATIVE
PACIFIC THEATER OF OPERATION

And the names of the great battles on the left side:

PEARL HARBOR – CORAL SEE –
CORREGIDOR – GUADALCANAL – TARAWA
– SAIPAN – LEYTE GULF – IWO JIMA –
OKINAWA – TOKYO

First World War commemorative

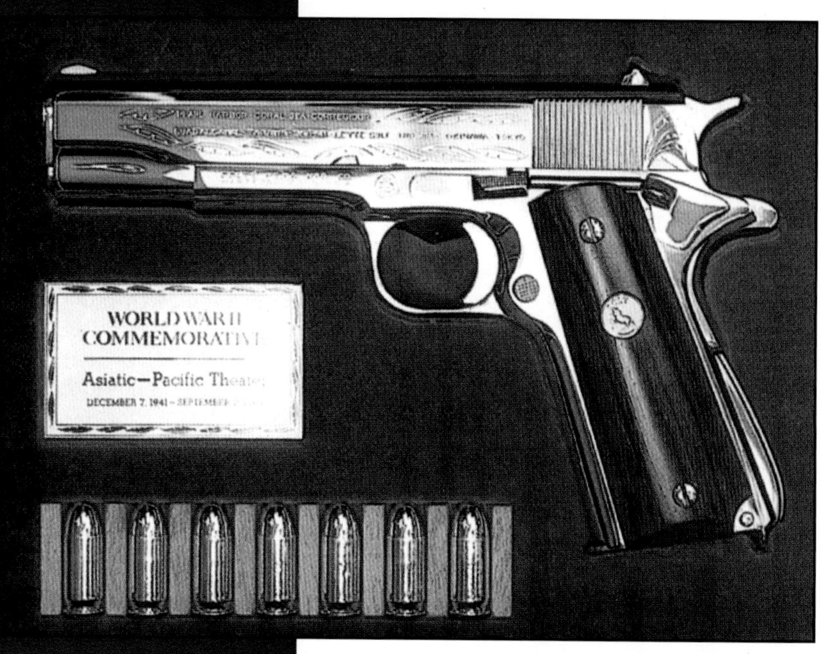

Pacific war commemorative

Vietnam commemorative Colt pistol

Officer of the 198th Infantry brigade in Vietnam, February 1965

The serial number is followed by *PTO*.

Every weapon is delivered in a luxury wooden box, with seven inert nickeled rounds, a map of the theater of operations, and a historical reminder of the combats.

Other commemorative weapons followed on a regular basis:

In 1971:

- The NRA Colt made 2,500 examples based on the Gold Cup National Match, to celebrate the centenary of the creation of the American Association of Amateur Firearms Users.

In 1978:

- a weapon for the 150th anniversary of the state of Ohio (250 weapons)
- a Service Model Ace "Arkansas" (Ace, 200 weapons)
- a Battleship Edition, USS *Texas* (500 weapons)

In 1979:

- The flight commemorative model for the 75th anniversary of the Wright Brothers' flight. This weapon is presented in nickeled finish and in a wooden box.
- Minnesota State Patrol, 1929–1979 (500 nickeled weapons)
- Louisiana, 175th anniversary (200 nickeled weapons)
- Los Angeles Police Department
- Panama Canal, 1870–1914 (200 weapons)

In 1980:

- Ohio President Special Edition (250 weapons)
- US Marine Corps (500 weapons)
- Houston Police Department
- Battleship Edition, USS *Arizona* (500 weapons)
- Camp Perry National Match, 1903–1978 (Gold Cup, 200 weapons)
- Combat Special Edition (Gold Cup, 500 weapons)
- Drug Enforcement Agency (901 weapons)
- Combat Companion (175 weapons)
- Olympic Ace (Ace, 200 weapons)
- Olympic Gold Cup (Gold Cup, 200 weapons)
- Combat Companion (20 weapons)

In 1981:

- Colt Signatures Series (500 weapons)

The Colt Delta Elite

The Colt Double Eagle

Marking on the barrel of a Colt Mark IV

- Alberta Diamond Jubilee (200 weapons)
- Hawaii Organization of Police Officers (100 weapons)
- International Shooters (Gold Cup, 300 weapons)
- Mini Special Edition (Gold Cup, 250 nickeled weapons)
- John M. Browning .45 Automatic (3,000 weapons)
- 250th Anniversary of Baltimore, 1732–1982 (200 weapons)
- US Customs Special Agent (425 weapons)
- Waterton Police Department Centennial Edition (200 weapons)
- OSS (200 weapons)
- Oklahoma Department of Correction (210 weapons)
- Vietnam Special Edition (200 weapons)
- Colt Signatures Series (Ace, 1,000 weapons)

In 1982:
- Mississippi Special Edition (200 weapons)
- BATF Limited Special Edition (15 weapons)
- BATF Deluxe Special Edition (48 weapons)
- Confederated Air Force (100 weapons)
- Los Angeles County Sheriff (450 weapons)
- San Diego Police Department (200 weapons)

In 1983:
- Confederated Air Force Silver Jubilee (silver coated, 50 weapons)

In 1984:
- Silver Star (stainless steel, 1,000 weapons)

In 1989:
- Joe Foss Limited Edition (2,500 weapons)

Colt Mark IV Series 70

In 1970, Colt modified its range of manufacture and proposed a series of pistols designated Mark IV / Series 70. These weapons are characterized by a new collet barrel bushing, designed to improve the centering of the tube, and the method of disassembly is slightly modified.

This new pistol is available in .45 ACP, .38 Super Auto, and 9 mm Parabellum, and it is available in multiple versions. The mains ones are

- Government Model
- Commander
- Combat Commander
- Gold Cup

It featured in the catalog until 1983.

Colt M1971

This is a prototype developed in 1971 from the Government Model. It is made from stainless steel, with grip plates in a synthetic material.

Colt M1911

Colt CO2, faithful copy of the original weapon, but a compressed air version

The encircling slide is guided hroughout its whole length. The veapon is operated by short barrel ecoil, but a sloping ramp replaced he connecting rod, whereas a protrusion at the mouth f the barrel takes the place of the barrel bushing.

The grip safety disappeared to make way for manual safety mounted at the rear of the slide. he trigger is single or double action.

The large-capacity magazine receives the ounds in a double-stack, double-feed system .45 ACP, .38 Super Auto, and 9 mm Parabellum).

This prototype was not developed, but some f its features are found on the SiG-Sauer P225.

olt Mark IV Series 80

ppearing in 1984, this new series is characterized y the security of the firing pin. It is presented in he same variations as the previous model.

olt Officer

lodel with an 89 mm barrel, developed in 1986.

olt Lightweight Officer

ersion of the previous weapon built with a light lloy frame.

olt Delta Elite

lew pistol presented from the series 80, firing 0 mm Auto rounds (1987).

olt Double Eagle Series 90

combat pistol entirely made of stainless steel, s silhouette is derived from the M1911. The rigger is single and double action. It is chambered n .45 ACP or 10 mm Auto.

olt Series 90

here is also a ninety range, which takes up the eatures of the previous ones, with a longer hook, hortened grip safety, grip in neoprene, adjustable ights, etc.

Colt M1911A1

In 1992, Colt relaunched the manufacture of the Government Model, rechristened the M1991A1. This was without doubt to respond to demand from amateurs and also to fight the competition, which was taking market share away with the many copies of the Government Model. Starting in 1994, it was presented in Compact and Commander versions, and a model in stainless steel was added. The numbering of these weapons is carried out following on from governmental supplies suspended in 1945.

Colt CO2

From 1996 onward, Colt proposed a range of compressed air pistols with the same appearance and identical dimensions as the Government Model. This model in 4.5 mm caliber operates with a capsule of carbonic gas. It has a slide fitted with oblique grooves.

It has a black or silver finish, with a choice of grip either in black plastic with grid pattern or smooth in plastic. In addition, there is another commemorative version to celebrate the 160th anniversary of the brand.

Detail of the markings on the Norwegian M1914 Colt. *Luc Guillou*

CHAPTER 9

COPIES AND DERIVATIVES

The Argentine Colt M1927 with its specific markings. *Luc Guillou*

Norway

Norway took up the Colt M1911 in 1914 in its original version, and several deliveries were carried out by Colt, using weapons taken from the commercial series:

- May 26, 1915, one pistol, no. C18158
- June 1, 1915, 400 others numbered from C18501 to C18850
- January 31, 1917, 300 weapons numbered from C88901 to 89200

All the pistols were imported by J. Mollbuch Thellefson.

Throughout 1917, Norway organized its own production at the arsenal at Kronsberg. The pistol is marked "COLT AUT. PISTOL M/1912" on the first one hundred examples. The weapon was then marked "11.25 m/m AUT. PISTOL M/1914."

In 1919, after the delivery of five hundred examples of a type similar to the American weapon, the model was modified. A slide stop with an extension was fitted, which allowed it to be handled with gloves. Kongsberg Vapenfabrik made more than 20,000 such pistols (between 21,941 and 22,311, depending on the source) up until 1940.

Production then continued under German control; the weapon bore the mark Pistole 657(n), and production is estimated at between 8,000 and 10,000.

Argentina

Argentina also received Colt M1911 pistols for the navy; they were taken from the commercial series but equipped with a specific marking (see chapter 3):

- 100 pistols on April 11, 1914, numbered from C6201 to C6300 by the intermediary of the River Ship Building Co., Quincy, Massachusetts
- 100 others on the same day, nos. C6301 to C6400, via Camden Ship Building Co., Camden, New Jersey
- last, 121 pistols on November 10, 1914, bearing serial numbers from C11501 to C13500

All the weapons feature the marking MARINA ARGENTINA on the right side of the slide.

The following year, on July 19, 1915, one housand pistols were delivered to the army nos. C20001 to C21000); because of the trade greements between Colt and the Fabrique Nationale (FN), it was the British agent of the FN hat delivered to the Argentinians.

Star A pistol

Super Star pistol

The Military Model is fitted with a selector.

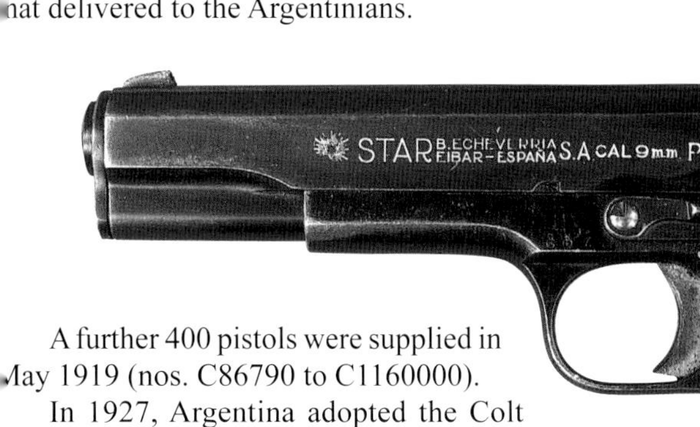

A further 400 pistols were supplied in May 1919 (nos. C86790 to C1160000).

In 1927, Argentina adopted the Colt M1911A1 pistol, and it ordered 10,000 from Colt with a specific marking: on the right side of the lide, the Argentinian coat of arms and the marking

EJÉRCITO ARGENTINO
COLT CAL. .45 MOD 1927

These weapons are numbered in a special series: from 1 to 10000.

After the Second World War, Colt conceded production license to Argentina, which was perated by the Dirección General de Fabricaciones Militares (DGFM), which ensured the manufacture t the Fábrica Militar de Armas Portatiles (FMAP) t Rosario in the province of Santa Fe.

These pistols are marked on the left side of he slide with

D.F.G.M. (F.M.A.P.)

And on the right side, with the Argentinian coat f arms and the marking

EJÉRCITO ARGENTINO
SIST COLT CAL. 11.25 mm MOD. 1927

They are numbered from 10001 to 112000. The rip plates are in black Bakelite.

Star

The Spanish weapons industry, concentrated ssentially in the Basque country, underwent a apid growth during the First World War, with onsiderable orders from the French army (more han a million pistols and revolvers) and to a lesser xtent, orders from Great Britain and Italy.

At the beginning of the 1920s, several manufacturers chose to rush into the gap left by Mauser to deliver weapons to a client with immense eeds: China. The majority of them (Astra, Beistegui, Hermanos, Eulogio Aristegui) chose o copy the Mauser C96 by modifying it and ometimes fitting it with a selector that permitted apid-burst fire.

Bonifacio Echeverria, which marketed weapons under the name Star, chose another direction. It was to reproduce the Colt M1911A1 by simplifying it. Even though the outline of the weapon remained the same, the method of production was different: the grip safety was removed, the slide stop and the optional safety for the hammer spring housing were no longer removable, and the back of the grip was changed to a grid pattern. Initially designated "Military Model," the weapon was offered in several calibers:

- 9 mm Bergmann: Bayard regulation in Spain under the name of 9 mm Largo, this ammunition is interchangeable with the .38 Super Auto (Model A).
- 9 mm Parabellum (Model B), used by the German army during the Second World War; this version bears the stamps of the Waffenamt and more often the marking F Patr 08 (for cartridge model 1908) at the front of the slide on the left side.
- 7.63 mm Mauser (Model M)
- .45 ACP (Model P)

Llama pistol

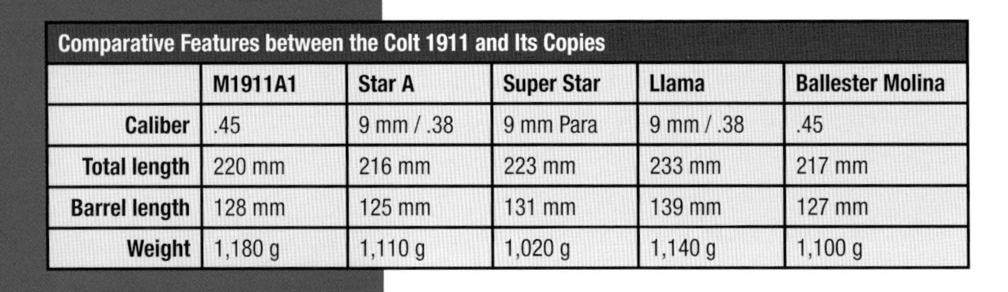

Comparative Features between the Colt 1911 and Its Copies					
	M1911A1	Star A	Super Star	Llama	Ballester Molina
Caliber	.45	9 mm / .38	9 mm Para	9 mm / .38	.45
Total length	220 mm	216 mm	223 mm	233 mm	217 mm
Barrel length	128 mm	125 mm	131 mm	139 mm	127 mm
Weight	1,180 g	1,110 g	1,020 g	1,140 g	1,100 g

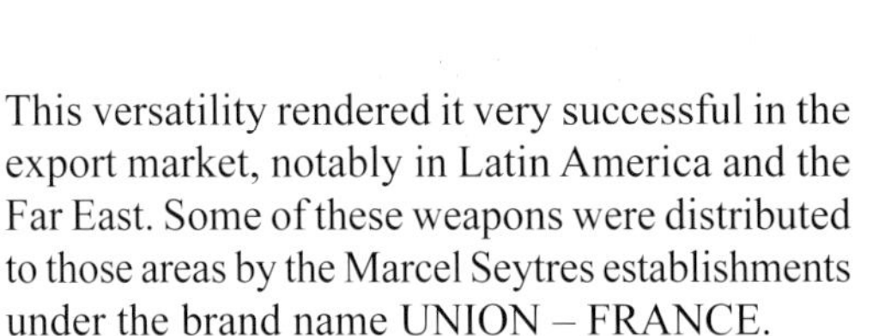

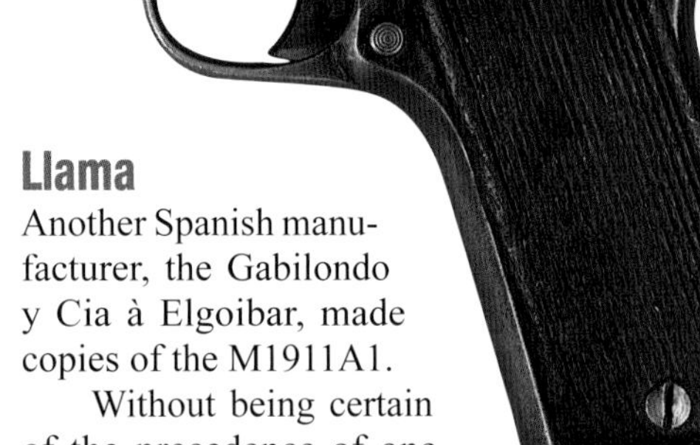

The Ballester Molina pistol was made with steel reclaimed from the wreck of the German pocket battleship *Graf Spee*.

This versatility rendered it very successful in the export market, notably in Latin America and the Far East. Some of these weapons were distributed to those areas by the Marcel Seytres establishments under the brand name UNION – FRANCE.

Some versions received a removable butt identical to that of the Mauser C96, a large-capacity magazine, and a fire selector.

Even though it had been used during the Spanish Civil War, the Star A became regulation in the Spanish army only from July 24, 1946 onward, and in the navy from October 25 in the same year.

Later, the outline of the weapon was slightly modified but the mechanism remained the same; the grip bulged at the base, which improved grip. The weapon was presented in versions AS and MS (9 mm / .38), BS (9 mm Para), MMS (7.63 mm), and PS (.45).

The Modelo Super appeared next; the barrel connecting rod disappeared to make room for a disassembly ramp, whereas disassembly no longer required the extraction of the slide stop: a lever positioned on the right permitted the unblocking of the barrel / slide unit. The extractor situated on the top serves as a loading indicator. This version seems to have been made only in 9 mm Parabellum (Super B), which was regulation army.

More recently, compact pistols made with a steel or light-alloy frame are seen from the same manufacturer, in 9 mm (BM and BKM) and .45 (PD). The conception of these pistols is moving further and further away from the original model.

Llama

Another Spanish manufacturer, the Gabilondo y Cia à Elgoibar, made copies of the M1911A1.

Without being certain of the precedence of one simplified copy of a Colt in relation to another, before the Second World War, Gabilondo marketed a versio of the weapon that was similar to the Star Militar Model. Their point in common is the absence o a grip safety on the back of the grip, and a percussion spring housing solid to the frame Conversely, Llama kept the common system o the maneuver spring for the slide stop and th optional safety, as on the Colt.

Heavier and bulkier than the Colt and the Sta this version was made in 9 mm Largo / .38 Supe Auto, and it was regulation in the Spanish army.

The weapons made subsequently were close to the original model; the grip safety, the hamme spring housing (which can be taken apart), an the barrel disassembly connecting rod all ar present. The manufacturer set up in Vitoria.

These pistols were made in numerous version

- 9 mm Largo (Model IX), also marketed under the name Tauler
- .45 ACP with a ventilation strip and a slight lip at the base of the grip (Model IX–A)

- .38 Super Auto, similar to the previous one (Model VIII)

- 9 mm Parabellum (Model XI)

Other weapons inspired from the Colt have been added to this list more recently; however, they are more compact or with a larger frame and an increased capacity, such as the .40-caliber S&W.

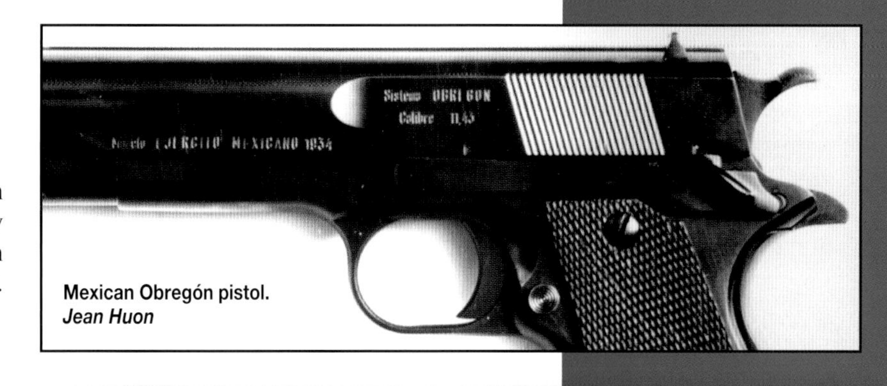

Mexican Obregón pistol. *Jean Huon*

Ballester Molina

In the thirties, a firm in Buenos Aires, Hispano-Argentina Fábrica de Automóviles Sociedad Anónima (HADFASA), made a simplified copy of the M1911A1; it is also very close to the Llama.

The weapon is chambered for the .45 ACP round, is characterized by the absence of a grip safety, the hammer spring housing is not removable, the trigger is pinned, and the mechanism is particular. The back of the grip is striated crosswise at the curved part, and the grip ridges of the slide are vertical and irregular. The grip plates are in wood with oblique vertical striations. The weapon is just a few millimeters smaller than the Model 1911A1, the magazines are interchangeable, and disassembly follows the same operation.

The Ballester Molina was used by the police and by certain paramilitary forces in Argentina. The weapons bore the marking POLICÍA FEDERAL. The example that has been examined was marked with the Argentinian coat of arms on the right side of the slide, followed by the marking POLICÍA MARITIMA; its serial number was greater than 28000.

During the Second World War, the British government ordered 10,000 Ballester Molina for its own needs. Because of successive delays, the order was not able to be delivered until 1943; at that time the needs of the British were less important, and they donated almost all the Ballester Molina to the Resistance organizations of western Europe, via the Special Operations Executive (SOE) and the Office of Strategic Services (OSS).

A rumor suggested that these pistols could have been made from the steel retrieved from the wreck of the German pocket battleship *Graf Spee*, scuttled in the bay of Rio de la Plata on December 18, 1939.

Obregón

On March 22, 1926, Mexico bought 198 Colt pistols of the future model 1911A1, with particular markings. This weapon, described in chapter 4, was supplied by the H. B. Dyas Company of Los Angeles, California.

Elsewhere, Alejandro Obregón from Mexico City developed a pistol conceived like the Colt M1911A1, but with a different locking system. A long lateral lever positioned on the left combined the safety and slide stop function.

General Motors experimental pistol in sheet metal. *KRCMlt*

The weapon was patented in the United States on April 26, 1938, under the number 2,115,041.

It was characterized by a helical-rotation locking system of the barrel; the recoil spring is mounted around the tube, and the slide has a protrusion on the front part.

The Obregón pistol was adopted by the Mexican armed forces and produced by the Fábrica Nacional de Armas de México.

General Motors

In 1945, with its wealth of experience acquired with the production of the single-shot Liberator FP-45, the General Motors company launched a study for a pistol based on the M1911A1, for the most part made from pressed sheet metal.

The war ended before the project was completed.

FNAB

At the beginning of the 1950s, the Fabricca Nazionale d'Armi de Bescia (FNAB) proposed a simplified copy of the Colt M1911A1 in 9 mm Parabellum to the Italian army. Eventually it was the Beretta M1951 that was retained.

Prototype of the Italian FNAB

The Springfield pistol

Detonics

In 1976, the Detonics firm was the first to make a clone of the Colt M1911A1; many others would follow after this, producing more or less faithful copies of the original.

The first model made by this company, based in Seattle, Washington, was an ultracompact model with an 89 mm barrel. Others followed, chambered for .45, .38 Super Auto, 9 mm Parabellum, or the very powerful .451 Detonics. These weapons present many improvements when compared with the original M1911A1 (backsight, spring, magazine, etc.).

Springfield Armory

In 1980, an American firm took over the company name of the Springfield Arsenal and became the Springfield Armory Inc. It initially made the M1A, a civilian version of the M14 (with machines bought from Beretta), then the Garand and finally M1911A1 pistols.

As with the majority of other producers, the basic model came in a multitude of versions, which at that point no longer had a lot to do with the original weapon.

Other Derivatives of the M1911A1 in the United States

Just like any successful object, whether weapons, watches, perfumes, sporting goods, or other consumer products, the M1911A1 has been and continues to be the object of numerous copies, derivatives, or even counterfeits. The following firms and establishments made such copies:

- AMT, Irwindale, California: HARDBALLER pistols and others in stainless steel with a longer barrel and slide
- Auto Ordnance, West Hueley, New York: exact copy of the Colt M1911A1 and numerous derivatives
- Coonan Arms, St. Paul, Minnesota: enlarged copies of the M1911A1, firing a .357 Magnum
- Drake Manufacturing Co.: produced frames with very tolerance for the assembly of Match pistols
- Essex Arms Corp., Island Pond, Vermont: made new frames for the commercial market

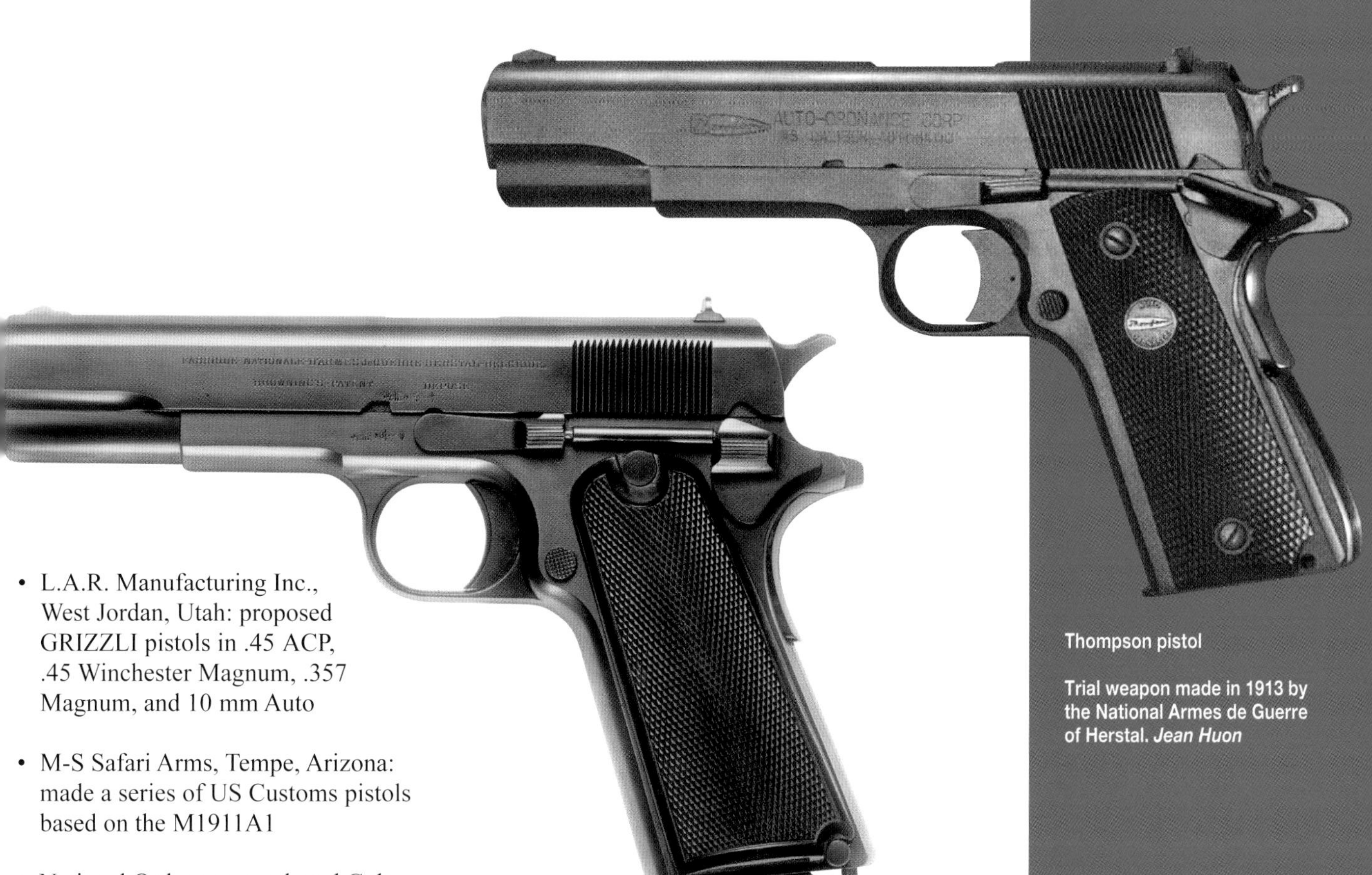

Thompson pistol

Trial weapon made in 1913 by the National Armes de Guerre of Herstal. *Jean Huon*

- L.A.R. Manufacturing Inc., West Jordan, Utah: proposed GRIZZLI pistols in .45 ACP, .45 Winchester Magnum, .357 Magnum, and 10 mm Auto
- M-S Safari Arms, Tempe, Arizona: made a series of US Customs pistols based on the M1911A1
- National Ordnance: marketed Colt M1911A1 frames after the war, bearing neither military stamps nor the marking U.S. PROPERTY. These were mounted from scrap, cut with a welding torch, and resoldered! On the right side, the marking NATL. ORD. CAL. .45 can be seen just above the serial number.
- Para-Ordnance, Scarborough, Ontario: proposed pistols derived from the M1911A1 with large-capacity magazines (thirteen rounds in .45)

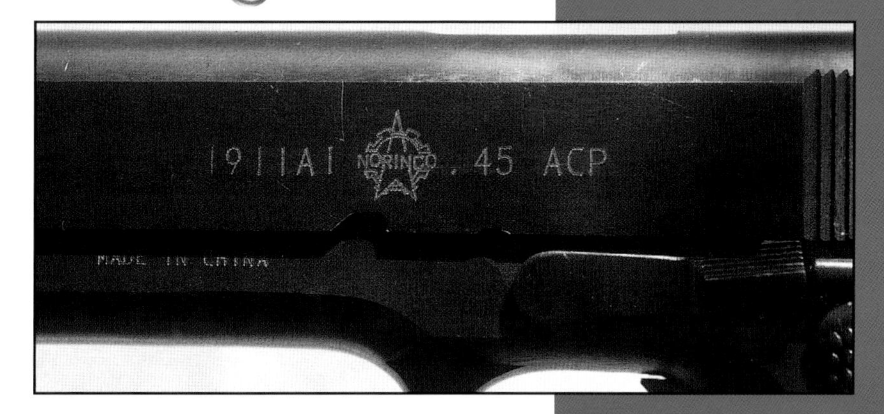

Markings on a Chinese copy of the Colt M1911A1 produced by Norinco

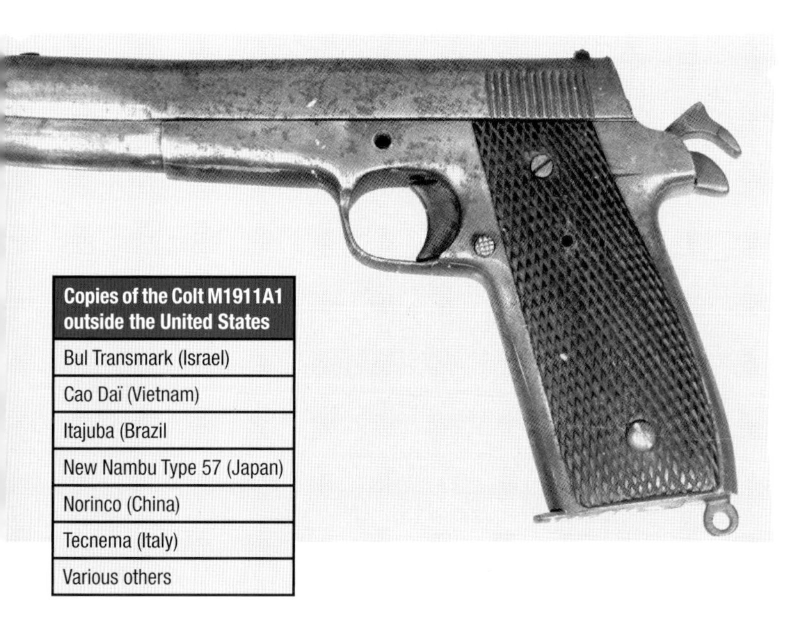

Copies of the Colt M1911A1 outside the United States
Bul Transmark (Israel)
Cao Daï (Vietnam)
Itajuba (Brazil
New Nambu Type 57 (Japan)
Norinco (China)
Tecnema (Italy)
Various others

Vietnamese copy of the Colt .45. *West Point Museum*

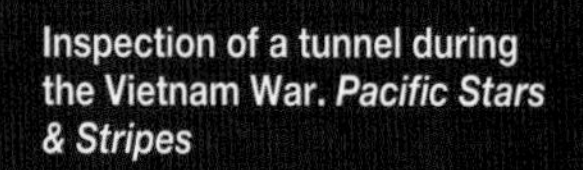

Inspection of a tunnel during the Vietnam War. *Pacific Stars & Stripes*

CHAPTER 10

ACCESSORIES

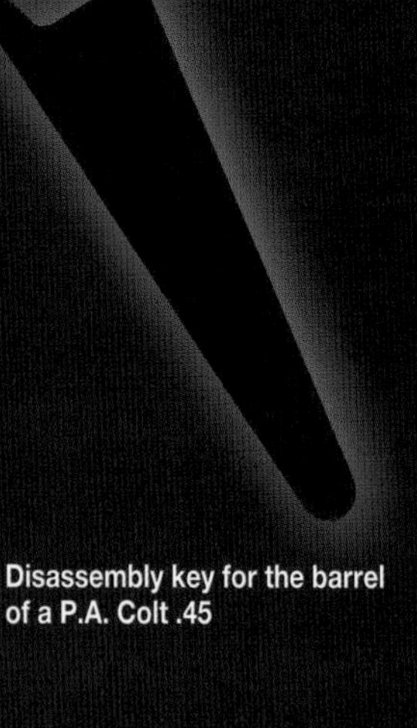

Disassembly key for the barrel of a P.A. Colt .45

Metal M1912 box containing maintenance accessories for the M1911 pistol. *Colin Doane*

Maintenance and Cleaning Accessories

A short time after the adoption of the Colt M1911, the Springfield Arsenal was given the task of creating maintenance equipment for the new pistol. Instead of assigning each weapon a group of accessories, these accessories were grouped in units.

That is how the Pistol Cleaning Kit M1912 came to be adopted, composed of a bronzed sheet-metal box with a handle, containing

- ten M4 brass cleaning rods, able to receive a chamber brush or a rag,
- ten M5 chamber brushes,
- ten tool units formed of small pieces of shaped, polished steel, bent at an angle and forming a pin remover extractor at one end and screwdriver at the other,
- a oil buret in brass, and
- a small box of grease, also in brass.

The same material was used during the second conflict; however, the bronzed surface treatment was replaced by a treatment with phosphate. The rods are in phosphate steel, as are the buret and the grease box. The tool unit is also treated with phosphate.

Private companies were also called on to produce some parts:

- Colt and Stanley for the tool unit
- PFAU Mfg. Corp., USCB Co., VHM, O'Hare Mfg. Co., J.A.M., and others for the rods, which are generally marked at the level of the ring

In addition, there is a key for disassembling the base of the barrel; this a nonregulation accessory but nonetheless is widely used among Colt .45 owners.

The box contained a screwdriver, cleaning rods, an oil buret, and a box of grease. *Marc de Fromont*

Models of oil burets and grease boxes. *Colin Doane*

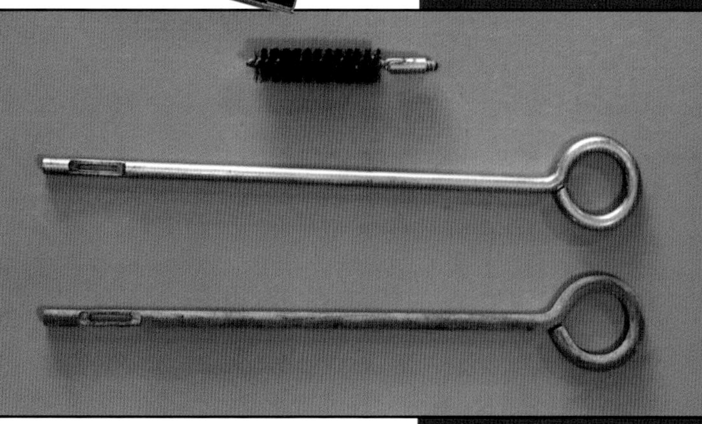

Cleaning rod in brass and steel and a chamber brush

Screwdriver

Pistol Rack

A wooden rack with six leather pouches (three on each side), with support straps.

This accessory was used by the US Navy, where, as in other navies of the world, the weapons were stored in the officers' mess . . . one could never be too sure!

Cartridge Case Trap

Small metallic cage adaptable for the Colt M1911 pistol and designed to catch ejected cases.

This was envisaged to be part of the kit for airplane crews using their service weapon during flight, so the ejected cartridge cases did not damage the fragile interior of the airplane.

A "cartridge case trap," this small cage was designed to retrieve the ejected cases of a Colt M1911 pistol, fired from an airplane, so that the interior of the plane was not damaged.

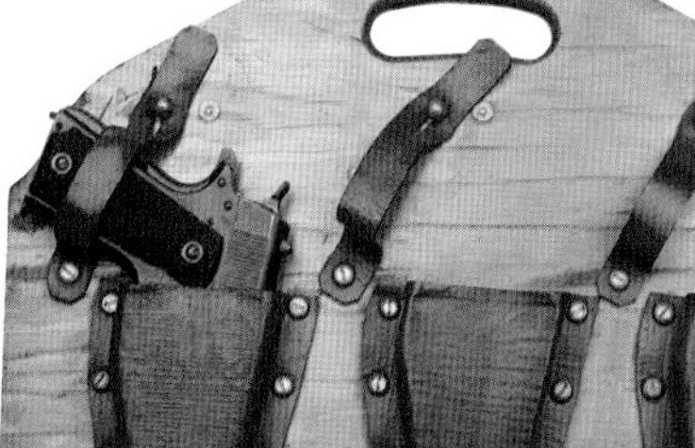

Navy pistol rack

Colt M1911A1 pistol with a twenty-five-round Spanish-origin magazine

Smooth Barrels for Lead Cartridges

In 1943, the ordnance tested several models of smooth or partially grooved barrels, of variable length (between 9.8 and 17.7 inches, or 25 and 45 cm). Made by Mossberg, these barrels were designed to enable airplane crews whose planes had landed in enemy territory to hunt for small game and thereby ensure their survival.

The 25 cm barrel was adopted in September 1945 but was never mass-produced, due to the end of hostilities.

Silencers

There were many sound reducers adaptable to the Colt M1911A1 in the form of a specific end piece of the barrel. They all were commercial productions, and their description falls outside the scope of this study.

Magazine with a twenty-five-round capacity, British made, with its transport pouch for two magazines. *Marc de Fromont*

Conversion from Pistol to Machine Pistol

At the beginning of the Second World War, a technician by the name of Franck Bialeski suggested a modification to the ordnance, permitting the transformation of the M1911A1 pistol into a small carbine capable of continuous-burst fire!

The weapon was fitted with an extended barrel, forming a flash concealer. It had a removable butt that could be used to house a twenty-round magazine.

The test report mentioned that this combination was less accurate than the M1 carbine, still undergoing assessment, and the project was abandoned in April 1942. The documents do not mention that this method was retained to fire the weapon in continuous burst.

Other inventors (Lebman, Swartz, and others) attempted similar modifications, but they did not result in mass production.

Large-Capacity Magazines

There are large-capacity magazines for Colt M1911 pistols, which have never been regulation in the army.

We have encountered

- twenty-round magazines of American origin,
- twenty-five-round magazines of American, British, or Canadian origin, and
- twenty-five-round magazines of Spanish origin conceived for the Star pistol and adaptable for the Colt M1911 or M1911A1.

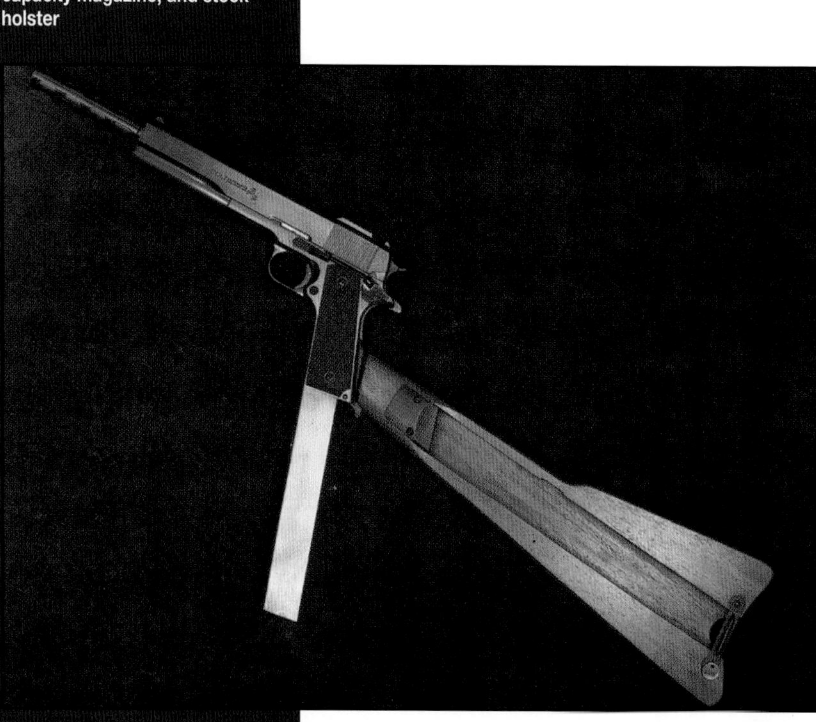

Experimental Colt pistol with fire mode selector, large-capacity magazine, and stock holster

CHAPTER 11

HOLSTERS

The use of the M1916 holster became more general with the M1911 pistol.

M1912 cavalry holsters and M1916 all-service holster. *Marc de Fromont*

As soon as the American army became interested in automatic pistols, it had holsters made for those models that were the most likely to be retained. From 1900 onward, therefore, small series of cases were made for Colt, Lüger, and Savage pistols.

M1912 Holsters for Mounted Troops

Starting in 1910, the ordnance researched a holster for a Colt pistol; from this first model came a holster initially specific to the cavalry but subsequently found in all units before another model was put into service: the Pistol Holster M1912.

Its conception combines both protection of the weapon and its rapidity of use, which leaves part of the grip visible. It is made of a pouch in thick, tawny leather, with a flap cut from the same piece. It is assembled with stitching reinforced with brass rivets. The inside has a block in wood covered with finer leather. At the bottom there is an eyelet to let any water that could get into the pouch drain away. The flap is closed by means of a brass stud and has a US monogram inside an oval.

A tab is fixed outside the pouch, and on the upper part of the tab is a large brass bearing. On the bearing is a swivel attachment with a metal hook for a US belt with eyelets. As a result of this setup, the pouch is always positioned vertically whatever the angle of the belt, which was of particular importance to the cavalry.

A strap placed halfway up the pouch allowed it to be fixed to the thigh of the user.

This holster was principally made by the State Arsenal of Rhode Island, although there were other manufacturers.

The Colt M1911A1 pistol and its M1916 holster, making an inseparable pair

The flap of the holster bears its characteristic US marking. *Marc de Fromont*

The holsters bear the manufacturer's mark on the inside, here the Rock Island Arsenal. *Marc de Fromont*

Marking specific to the firm G&K. *Marc de Fromont*

Alsace 1944. Consultation between French and American officers before the attack. The American major wears a Colt .45 in an M1916 holster, as well as an M3 dagger in its sheath.

M1912 Holster for Nonmounted Troops

With an organization similar to the previous model, this one is characterized by the presence of a nonjointed metal hook. The thigh strap was no longer a feature.

Made initially at the request of the Signal Corps, it was also used in other services.

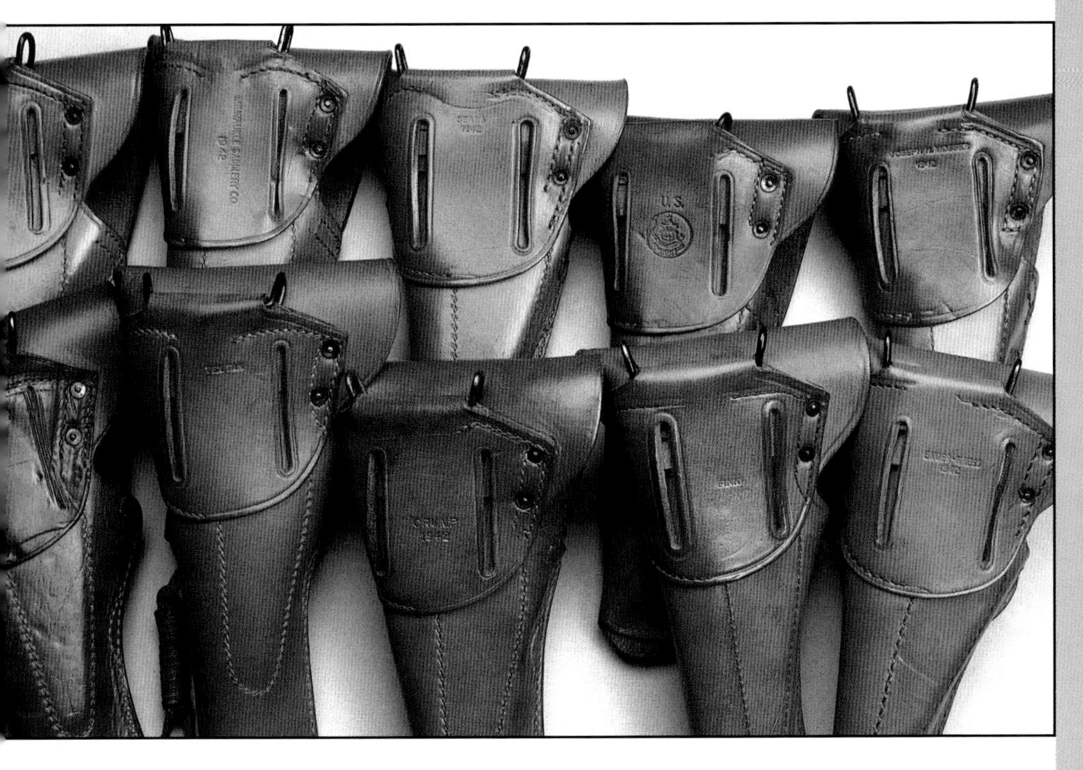

Sample of M1916 holsters. *Marc de Fromont*

Marking on a holster produced by Graton & Knight Co. *Marc de Fromont*

M1916 Holster

This holster is a simplified and more functional evolution of the 1912 model. The pouch, its internal block, and the flap all remain the same. At the lower part, two eyelets are fitted to allow any water to drain away; they also permit a leather lace to be inserted in order to fix the weapon to the thigh of the user if required.

The external part for suspension is of smaller dimensions, which brings the suspension hook to the level of the top of the flap and positions the weapon at a more convenient height for a rapid release. In addition, some modifications prevent the holster from moving around too much. Two notches in the suspension part also allow it to be carried higher on a standard belt.

Made initially from tawny, then black, leather, this holster was produced continuously from 1916 to 1986. Its numerous manufacturers and their company name appear, in theory, on the suspension part. They include, among others,

- Rock Island Arsenal
- A.L.P. Co.
- Bannerman
- Bolen Leather Products
- Boyt
- Brauer Bros. Mfg.
- Cathey Enterprises
- Clinton
- Craighead
- G&K
- G.P.&S
- K.B.: H.A.B
- Perkins & Campbell

There are variations in white leather, in polished black leather, and models close to the original made in higher-quality leather by saddlers.

Mills Holster

Designed for the US Marines, the Mills holster was subject to a restricted circulation during the First World War.

Made from a single piece in a strong olive-green canvas, it is characterized by a flap that was more enveloping than that of the army model. It closed by means of a press stud. An additional tab allowed the mounting of a suspension hook and a brass snap is the end of the pouch.

M1 Holster in Canvas

This was an experimental model made in 1934. An attempt was made to make a pouch in strong canvas for the Colt M1911A1 pistol on the same model as the M1916, but the tests did not lead to a satisfactory result.

In 1942, an identical attempt was made to make a holster in rubberized canvas.

Holsters for Officers

From the Second World War onward, the generals of the American army used holsters in black or tawny leather of a higher quality, enabling the

A para-officer in Algeria. His TAP1 holster could carry a Colt .45, P.08, P38, or PA50.

M3 shoulder holsters. *Marc de Fromont*

regulation Colt M1911A1 to be housed, or its short-barrel variations (Commander or M15).

These holsters either have a flap or a support tab. They are designed to be worn attached to a leather belt by means of a loop.

M3 Shoulder Holster

In 1942, at the request of aviators, an open holster was made and designed to be worn at chest height by means of a diagonal strap. A tab held the weapon in position, and a loop at the lower part meant the case was connected to the belt.

This unit is made from tawny leather and was also used by land troops, notably armored tank crews.

Officers in a discussion on an air base in Vietnam. The one in the center wears an M7 holster.

M7 Shoulder Holster

A variation of the previous model, it is distinguishable solely by its method of fastening: a sling combined with a belt that goes around the chest and makes it less mobile. The case remains unchanged.

This model, appearing in 1944, is made with tawny or black leather.

M9 Holster

Reviewed at the end of the Second World War, this holster, initially designated M7E3, is a modification of the previous model. The case is made of rubberized canvas and is the same shape as the M7. The harness has slings that cross over the back. Adopted in April 1945 under the name M9, this holster was never mass produced.

M66 Holster

The M66 appeared in 1970. Designed and made by Bianchi, it would seem that it was used only by military police personnel. It has a case in black leather with a removable, jointed flap with two press studs, which facilitate a quick release.

It is perfectly ambidextrous, with a loop at midheight on either side of the case, which means that the weapon is placed fairly high.

M12 Holster

A holster made by Bianchi or Cathey, in khaki, black, or camouflage nylon. There are two sizes, enabling the majority of automatic combat pistols, most notably the Colt M1911A1 or its successor the Beretta 92 (known as M9 in the US), to be housed. It is attached to the belt by means of sliding clips.

Close-up of M3 holsters. *Marc de Fromont*

British Holster

The British (commandos or parachutists) used a holster of a similar design to those for Enfield and Webley revolvers. The holster was very enveloping and was made of khaki canvas with a flap that closed with a press stud.

M66 Bianchi holster is ambidextrous, reversible, and able to be used either with or without the flap.

All-purpose M12 holster.

Brazilian Holster

A flexible holster in green canvas, closing by means of a flap with an American elastic metal button. A housing for a spare magazine, with a flap with a press stud, is placed on the external side.

An NCO radio operator in Vietnam. He is wearing a Colt .45 in a holster of local manufacture, placed very low on the thigh.

French Holsters

A triangular leather holster was made for the Colt M1911 used by tank crew in 1915; this was a larger version of the Ruby holster.

During the Second World War the French used M1916 American holsters, which remained in service in Indochina and Algeria. But the Colt could also fit into G.T.M.48 leather "hold-all" holsters, TAP1 or 2 in canvas, which could hold the PA50, Colt, P38, or P.08 with varying degrees of fit.

During the Korean campaign, there were locally made holsters in existence that were fairly faithful copies of the M1916 holster.

Magazine Pouches

The Colt M1911 was initially put into service with a leather pouch with a flap with press stud. It could hold two pistol magazines.

Leather was swiftly abandoned, however, for fabric, and the Mills magazine holder was adopted in 1914. Made from a strong khaki canvas, it had two pockets for magazines, its flap closed by means of two press studs. On the first models these studs bore the American eagle, but this practice was later abandoned. Each magazine pouch was accompanied by a 3-by-4-inch piece of paper warning the user of the dangers of damp and rust. A loop on the back allowed the pouch to be put on a belt.

The manufacture of this pouch was entrusted to other companies during the First World War: F.S.F., L. C. Chase Co., R. H. Long, Plant Brothers Co., etc. The press studs were replaced by oval metallic buttons, closing with two elastic tongues. The shape of the flap was sometimes redesigned, and a support press stud was added on the belt.

The magazine pouch evolved slightly with the arrival of the Second World War. Its canvas became olive green, and a single, central button was used to close the flap. Other manufacturers were called on: Avery, Ero Mfg. Co., E.O. Inc., Hoff Mfg. Co., and R.M. Co.

After the war the loop was replaced by sliding clips, which became more general when the M14 magazine pouches were put into service.

Two models of French holster, with a back view of the first model, for Colt M1911, used by the crews of the first assault tanks. *Marc de Fromont*

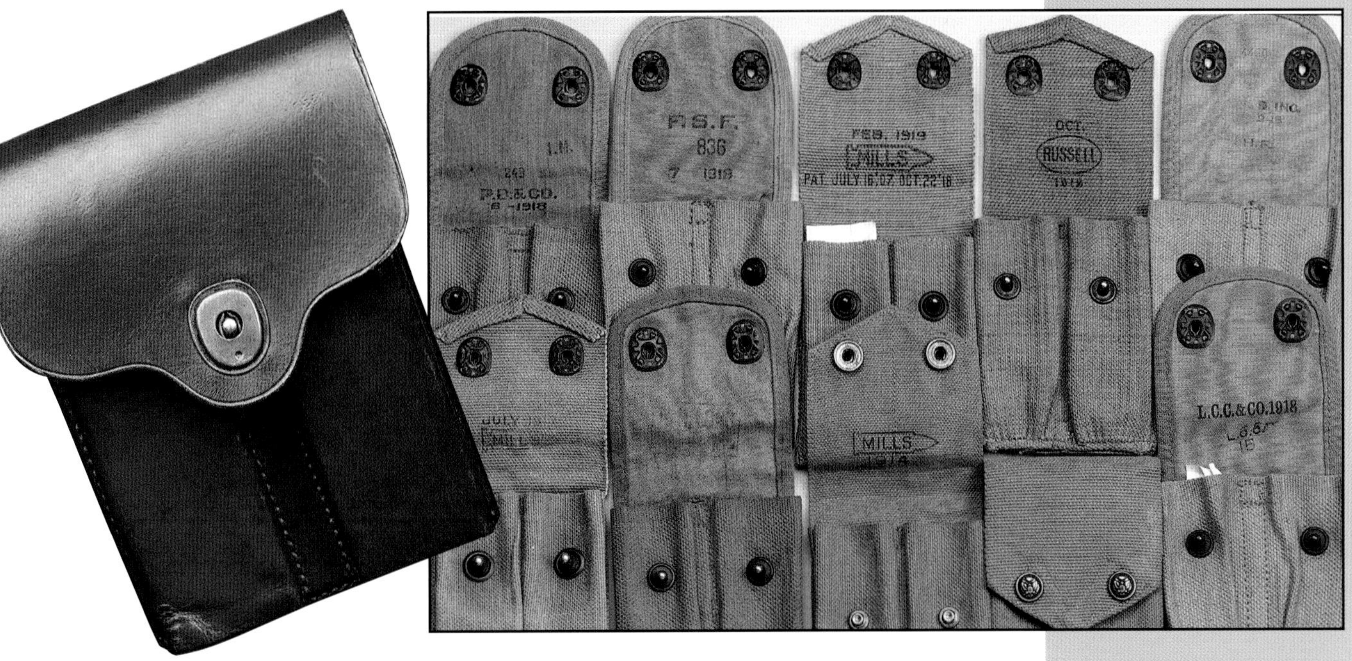

Leather pouch for Colt pistol magazines. *Marc de Fromont*

Range of magazine pouches produced during the Second World War. *Marc de Fromont*

Mills magazine pouch (ON LEFT, First World War style). *Colin Doane*

Magazine pouch (ON RIGHT, Second World War style). *Colin Doane*

There are magazine pouches in blue canvas for the US Air Force and white canvas for parades. There are also ones in brown leather, made from high-quality leather by saddlers, for high-ranking officers or generals. The flap is closed by means of one or two press studs.

Lanyards

In 1917, a lanyard was put into service for the Colt M1911, consisting of a double lace in khaki with a leather tightening loop and a hook at the end.

The lanyard used during the Second World War is basically similar. It is an olive-green color, and its hook is mounted on a leather tab.

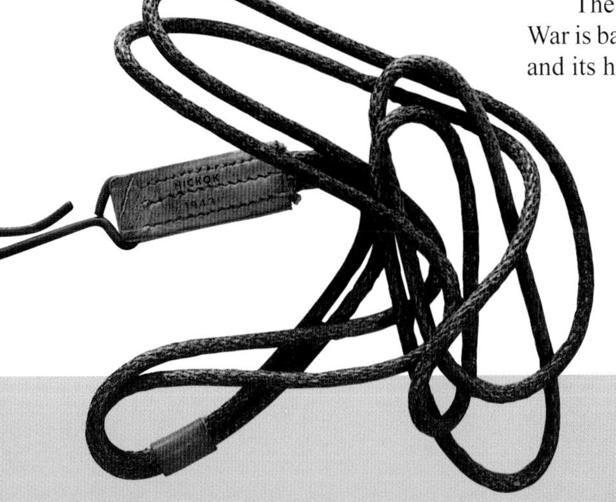

Lanyard in service during the Second World War. *Colin Doane*

No training without ammunition—this meant that in general, soon-to-be out-of-date stocks could be used up. *US National Archives*

CHAPTER 12

AMMUNITION

.45 ACP cartridges in compartmentalized packaging. *Marc de Fromont*

.45 ACP cartridges placed top to tail in a box. *Marc de Fromont*

The first research relating to a military cartridge for a .45 automatic pistol began in 1904 at the Frankford Arsenal, in northeast Philadelphia. The Winchester company was also interested following a request from Colt.

In 1906, Frankford made the first .45 ammunition for the American army. Its outline and characteristics were already very similar to the cartridge that was to become the .45 ACP.

The .45 ACP 1911 Model Cartridge

The following year, the Union Metallic Cartridge Co. received an order from the government for a new version of the cartridge with a deeper extraction groove. The definitive ammunition was finalized in 1908–1909.

Adopted in 1911, it consisted of a grooved case in brass, with its cylindrical surface receiving a smokeless powder charge. Cases were also made in chromate steel. The cylindrical hemispheric bullet was made of a core of lead and a tombac case; its mass was 14.9 g. There was sometimes a groove at midheight, which facilitated loading by limiting the recess of the projectile.

The initial speed of the projectile is approximately 800 feet (244 m) per second in the barrel of the pistol, which gives it 443 joules (45 kgm) of energy. At 14.76 feet (4.5 m) from the mouth of the barrel, it can penetrate six 1-inch-thick planks of pine.

Variations

- Tracer bullet cartridge US M1, US T30, and US M26 (red point)

Different boxes of American ACP cartridges: pistol ball, lead, and dummy. *Marc de Fromont*

Characteristics	
Bullet diameter	11.43 mm
Diameter of case at neck	11.92 mm
Diameter of case at rim	11.94 mm
Diameter of case at base	12.05 mm
Length of case	22.7 mm
Total length	31.8 mm
Projectile mass	14.9 g

TOP, commercial cartridges. Characteristic wrapping of American ammunition of the Second World War. *Col. A. Nowak*

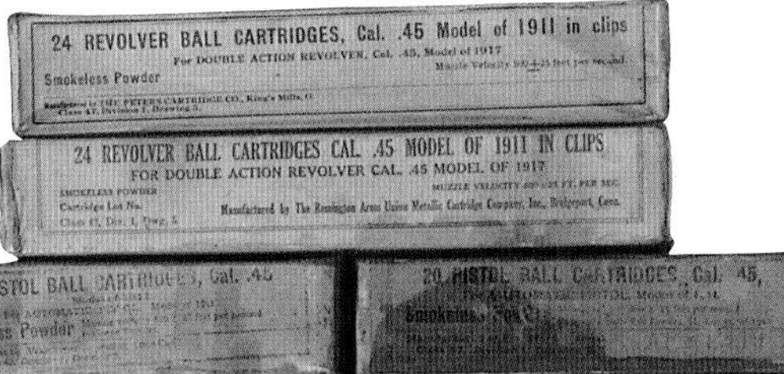

Boxes of ammunition with First World War wrapping. *Col. A. Nowak*

Firing exercise by American soldiers arriving in France in 1917. Note the empty shell during the ejection phase. *Jean-Claude Kempf*

Soldiers of the 60th Infantry Regiment (9th Division) during a patrol as part of operation "Hot Tac" in Vietnam in April 1967. The man in second position is carrying a Colt M1911A1 and a M3A1 submachine gun.

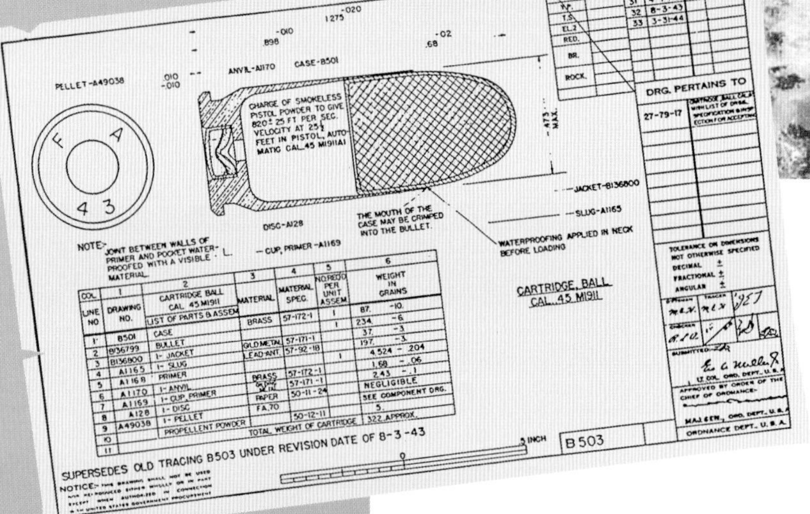

PELLET-A49038

ANVIL-A1170 CASE-B501

CHARGE OF SMOKELESS PISTOL POWDER TO GIVE 820± 25 FT PER SEC. VELOCITY AT 25½ FEET IN PISTOL, AUTOMATIC CAL .45 M1911A1

JACKET-B136800

SLUG-A1165

DISC-A128

THE MOUTH OF THE CASE MAY BE CRIMPED INTO THE BULLET.

WATERPROOFING APPLIED IN NECK BEFORE LOADING

NOTE: JOINT BETWEEN WALLS OF PRIMER AND POCKET WATERPROOFED WITH A VISIBLE MATERIAL.

CUP, PRIMER -A1169

CARTRIDGE, BALL CAL. .45 M1911

COL.	1	2	3	4	5	6
LINE NO	DRAWING NO.	CARTRIDGE BALL CAL .45 M1911 LIST OF PARTS & ASSEM	MATERIAL	MATERIAL SPEC.	NO. REQ'D PER UNIT ASSEM	WEIGHT IN GRAINS
1	B501	CASE	BRASS	57-172-1	1	87. -10.
2	B136799	BULLET			1	234 -6
3	B136800	1- JACKET	GLD METAL	57-171-1		37. -3
4	A1165	1- SLUG	LEAD-ANT	57-92-18		197. -3
5	A1168	PRIMER			1	4.524 - .204
6	A1170	1- ANVIL	BRASS	57-172-1		1.68 - .06
7	A1169	1- CUP, PRIMER	GILDING METAL	57-171-1		2.43 - .1
8	A128	1- DISC	PAPER	50-11-24		NEGLIGIBLE
9	A49038	1- PELLET	FA 70			SEE COMPONENT DRG.
10		PROPELLENT POWDER		50-12-11		5.
11			TOTAL WEIGHT OF CARTRIDGE			322 APPROX.

SUPERSEDES OLD TRACING B503 UNDER REVISION DATE OF 8-3-43

B503

Diagram of a standard M1911 ball cartridge

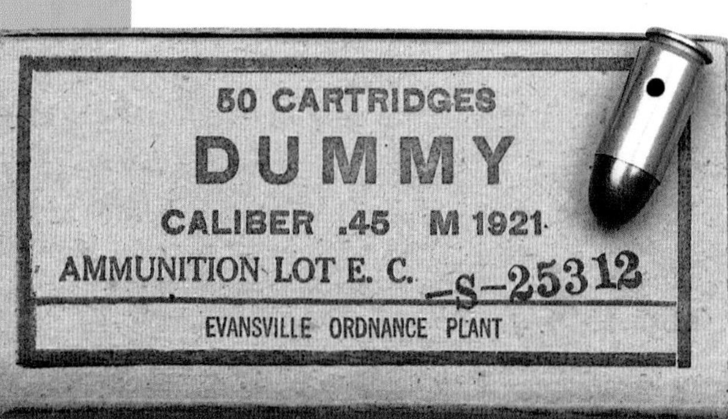

A box of M1921 dummy cartridges

- Survival cartridge US M12. A cylindrical hemispheric envelope in paper capsule containing an 11 g lead charge no. 7½.
- Survival cartridge US M15 Case lengthened to 31.8 mm and sealed with a paper capsule disc; it encloses 108 to 118 lead no. 7½.
- Signal cartridge US T92. The burning powder composition is covered with a resin cap in the shape of a hemisphere.
- Blank cartridge US M9, case in steel chromate lengthened to 28.3 mm; the lips are tightened and rimless.
- Test cartridge US M1, case in galvanized brass; the bullet and base are covered in red varnish.
- Dummy cartridge US M1921; case pierced with one or two holes. Nonprimed.
- Dummy cartridge; case in chromate with empty primer cup, bullet with a steel jacket

The .45 ACP cartridge remains very widespread throughout the world and there are many variations; its manufacture was taken up by the majority of cartridge factories.

Ammunition for Colt .45: standard ball M1911 cartridge, survival cartridge, M11 lead. Lead M15 survival cartridge. Blank M9 cartridge and M1921 dummy cartridge.

This photo shows the Colt pistol "family tree," with (FROM TOP TO BOTTOM), Colt 1900, Colt 1911 made by Colt, another of the same model (Colt 1911) made by Remington, an M1911 chambered in .455 for the British, and the M1911A1. All are pictured with their accessories: holster, disassembly tool, rod, lanyard, cartridges, and oil buret. *Col. A. Nowak / Marc de Fromont*

EPILOGUE

The Colt M1911 inspired numerous pistols, among them the Tokarev TT30 and TT33, Browning GP35, Radom VIS35, French pistols 35A and 35S models, the SIG P210, and many others.

After the Vietnam War, the aging of the stock of Colt pistols, along with the multitude of .38-caliber revolvers that had been added to it, led the military to once again look for a successor to the M1911A1 and its variations.

A great number of American and European models of pistols were tested; the result was that the Beretta, presented by the American subsidiary of the Italian firm, and the SIG-Sauer, presented by Maremont, were left in the running.

It was finally the Beretta 92 SB-F that was retained and adopted in January 1982, under the name Pistol M9, to equip the US Army, Navy, Air Force, Marine Corps, and Coast Guard.

The Colt M1911A1 Today

It remains extremely popular despite its over 100 years of existence. Although its weight and single-action trigger are sometimes considered a handicap, it is nonetheless a safe, robust, powerful, and reliable weapon.

The numerous copies of the Colt that are produced almost everywhere in the world are proof of the admiration that still exists for this model.

BIBLIOGRAPHY

Periodicals

American Rifleman
Cibles
Guillaume Tell
Gun Digest
La Gazette des Armes
Man at Arms
Militaria
Profile Publications
Waffen Digest

Automatic Pistols M1911 and M1911A, US Army, FM-23-25 French, 1943.

Brandt, Jakob H. *Handbuch der Pistolen- und Revolver-Patronen*. Schwäbisch Hall, Germany: Journal-Verlag Schwend, 1998.

Cadiou, Yves L. *Les pistolets automatiques Colt*. La Tour-du-Pin, France: Éditions du Portail, 1996.

Canfield, Bruce N. *U.S. Infantry Weapons of World War II*. Lincoln, RI: Andrew Mowbray, 1994.

Canfield, Bruce N. "The U.S. '.45 Automatics.'" *American Rifleman*, November 2006.

Caranta, Raymond. *Pistolets et revolvers, aujourd'hui*. 2 vols. Chaumont, France: Crépin-Leblond, 1998–1999.

Caranta, Raymond, and Pierre Cantegrit. *L'aristocratie du pistolet*. Montrouge, France: Crépin-Leblond, 1997.

Catalogue of Light Arms and Carriages in Service in the Army Mat, 1080, Direction Centrale du Matériel, 1958.

Catalogue of Light Arms and Carriages in Service in the Army Mat, 1181, Direction Centrale du Matériel, 1982.

Clawson, Charles W. *Colt .45 Service Pistols: Models of 1911 and 1911A1: Complete Military History, Development, and Production 1900 through 1945*. Fort Wayne, IN: C. W. Clawson, 1983.

Cormack, A. J. R. *Small Arms Profile 5: The Colt .45 and Its History*. Windsor, UK: Profile Publications, 1973.

Crowell, Benedict. *America's Munitions, 1917–1918*. Washington, DC: US Government Printing Office, 1919.

École d'Application de l'Infanterie. *Notice sur le pistolet automatique US Mle 1911 A1*. Auvours, France: École d'Application de l'Infanterie, 1948.

État militaire de toutes les nations du monde. Paris: Berger-Levrault, 1914.

Ezell, Edward C. *Handguns of the World: A Comprehensive International Guide to Military Revolvers and Self-Loaders*. London: Arms & Amour, 1981.

Hacker, Larry. *The Colt 1911 Automatic Pistol: Its Predecessor and Variations; A Pocket Chronology*. Little Rock, AR: Larry Hacker, 1983.

Hackley, Frank W., William H. Woodin, and Eugene L. Scranton. *History of Modern U.S. Military Small Arms Ammunition*. 2 vols. Rev. ed. Gettysburg, PA: Thomas, 1998.

Hallock, Kenneth R. *Hallock's .45 Auto Handbook*. Oklahoma City, OK: Mihan, 1980.

Hartnik, A. E. *Pistolen en revolvers encyclopedie*. Lisse, The Netherlands: Rebo, 1996.

Hoffschmidt, E. J. *Know Your .45 Auto Models 1911 and A1*. Chino Valley, AZ: Blacksmith, 1974.

Hoffschmidt, E. J., and Jean Huon. *Les pistolets Colt: Modèle 1911, modèle 1911 A1*. Versailles, France: Éditions Sofarme, 1979.

Hughes, James B., Jr. *Mexican Military Arms: The Cartridge Period, 1866–1967*. Houston, TX: Deep River Armory, 1968.

Huon, Jean. *Un siècle d'armement mondial: Armes à feu d'infanterie de petit calibre*. 4 vols. Paris: Crépin-Leblond, 1976–1981.

Huon, Jean. *Les pistolets automatiques français, 1890–1990*. Paris: Histoire & Collections, 1995.

Johnson, George B., and Hans B. Lockhoven. *International Armament: With History, Data, Technical Information and Photographs of over 400 Weapons*. Cologne: International Small Arms Publishers, 1964.

Leyson, Burr W. *The Modern Colt Guide*. New York: Greenberg, 1953.

Malherbe, Michel. *Le Colt .45 Auto*. Verneuil-sur-Avre, France: F. G. Éditions, 1987.

Meadows, Edward Scott. *U.S. Military Holsters and Pistol Cartridge Boxes*. Dallas: Taylor, 1987.

Medlin, Eugene, and Jean Huon. *Military Handguns of France, 1858–1958*. Latham, NY: Excalibur, 1993.

Ministère de la Défense National. *Manuel du gradé, partie commune a toutes les armes*. TTA 116. Paris: Charles-Lavauzelle 1956.

Montrevel, Sébastien. *Le Colt .45*. Paris: Éditions du Guépard, 1981.

Parra, Numa-Marie. *Pistolets automatiques*. Paris: Berger-Levrault, 1899.

Royer, Jean-Pierre-Joseph, and Bernard-Charles-Eugène Millet. *Recueil de résumés d'instruction militaire élémentaire*. Paris: La Renaissance, 1964.

Smith, W. H. B., and Joseph E. Smith. *Book of Pistols and Revolvers*. 7th ed. Harrisburg, PA: Stackpole Books, 1968.

Still, Jan C. *Axis Pistols: The Pistols of Germany and Her Allies in Two World Wars*. 2 vols. Marceline, MO: Walsworth, 1986.

US Army Ordnance Department. *Record of Army Ordnance*. Vol. 2. Washington, DC: Office of the Chief of Ordnance, Research and Development Service, 1946.

US Army and US Air Force. *Field Maintenance, Cal. .45 Automatic Pistols M1911 and M1911A1*. Washington, DC: US Army and US Air Force, 1957.

US Department of the Army. *Pistol, Caliber .45, Automatic, M1911A1*. Technical Manual 9-1005-211-34. Washington, DC: Headquarters, US Department of the Army, 1964.

US Department of the Army. *Pistol, Caliber .45, Automatic, M1911A1, with Holster, Hip (1005-673-7965), and Pistol, Caliber .45, Automatic, M1911A1, with Holster, Shoulder (1005-561-2003): Operator and Organizational Maintenance Manual*. Technical Manual 9-1005-211-35. Washington, DC: US Department of the Army, 1968.

US Department of the Army. *Pistols and Revolvers*. Field Manual 23–25. Washington, DC: Headquarters, Department of the Army, 1971.

Wilson, R. L. *The Colt Heritage: The Official History of Colt Firearms, from 1836 to the Present*. London: Jane's, 1979.

Thanks to

Gaétan Brunel, Raymond Caranta, Jean-Louis Courtois, Jean-Claude Dey, Colin Doane, Jean-Claude Kempf, Oseph Lapicirella, Gene Medlin, André Nowak, Alain Tomei, and Bill Woodin. All uncredited photos are courtesy of Jean Huon.

CLASSIC GUNS OF THE WORLD SERIES